Cooper, David J.; Merritt, David M. 2012. **Assessing the water needs of riparian and wet-land vegetation in the western United States**. Gen. Tech. Rep. RMRS-GTR-282. Fort Collins, CO: U.S. Department of Agriculture, Forest Service, Rocky Mountain Research Station. 125 p.

Abstract

Wetlands and riparian areas are unique landscape elements that perform a disproportionate role in landscape functioning relative to their aerial extent on the landscape. The purpose of this guide is to provide a general foundation for the reader in several interrelated disciplines for the purpose of enabling him/her to characterize and quantify the water needs of riparian and wetland vegetation. Topics discussed are wetland and riparian classification, character-istics and ecology, surface and groundwater hydrology, plant physiology and population and community ecology, and techniques for linking attributes of vegetation to patterns of surface and groundwater and soil moisture.

Keywords: riparian, wetland, groundwater, water requirements, vegetation

Authors

David J. Cooper, Department of Forest, Rangeland and Watershed Stewardship, Colorado State University, Fort Collins.

David M. Merritt, National Watershed, Fish, Wildlife, Air, and Rare Plants Staff, USDA Forest Service and Natural Resource Ecology Laboratory, Colorado State University, Fort Collins.

You may order additional copies of this publication by sending your mailing information in label form through one of the following media. Please specify the publication title and number.

Publishing Services

Telephone	(970) 498-1392
FAX	(970) 498-1122
E-mail	rschneider@fs.fed.us
Web site	http://www.fs.fed.us/rmrs
Mailing Address	Publications Distribution Rocky Mountain Research Station 240 West Prospect Road Fort Collins, CO 80526

Contents

Chapter 1: Introduction

Wetland and riparian ecosystems comprise a very small percentage of the western U.S. land area, yet provide important economic and ecological functions (Gregory and others 1991, Patten 1998, Mitsch and Gosselink 2000). Wetlands provide important habitat for many species of animals, particularly amphibians, birds, and mammals (Nelson and others 1984, Haukos 1992, Brown and others 1996, Davidson and Knight 2001); are local and regional centers of biodiversity (Naiman and others 1993, Pollock 1998, Nilsson and Svedmark 2002); and provide biogeochemical, physical, and ecological processes that maintain water quality, flood attenuation, forage production for livestock, watershed hydrologic functioning, stream and lakeside stability, and a range of other valuable ecosystem services.

As the ecological importance of wetland and riparian ecosystems has become better understood, laws and regulations have been promulgated toward ecosystem conservation and management (e.g., Clean Water Act, National Environmental Policy Act, Endangered Species Act, and floodplain regulation). However, many wetland and riparian ecosystems in the United States have been damaged or destroyed by anthropogenic activities, including drainage for agriculture, dewatering and altered flow regimes by dams and reservoirs and groundwater pumping, stream water diversions, filling, gravel mining, and other activities (Tiner 1984, Patten 1998, Graf 1999, Brinson and Malvarez 2002). Of growing concern is the increasing human demand for water, particularly in arid and semiarid regions of the West. This demand is intensifying the pressure on rivers and their adjacent riparian areas, wetlands, and groundwater systems and is threatening the functioning and long-term viability of these areas (Pringle 2000, Baron and others 2002).

Purpose and scope

The U.S. Department of Agriculture Forest Service manages 193 million acres of National Forests and Grasslands in the United States, which includes over 400,000 miles of streams and rivers and 3 million acres of lakes. The National Forests are the largest single source of drinking water in the United States, providing 20 percent of the nation's water supply. The Forest Service is responsible for balancing often conflicting multiple uses when managing public lands for "favorable conditions of flows" while enhancing the quality of life for the American public by supporting agriculture; sustaining the health, diversity, and productivity of the nation's forests and grasslands; and supporting recreation, mining, timber harvest, energy development, and water extraction. The estimated value of water and ecosystem services provided by healthy National Forest and Grassland watersheds currently exceeds $7 billion annually (Brown 2004). Most U.S. Forest Service land management plans highlight the economic and ecological importance of maintaining the biological integrity of aquatic and riparian ecosystems. Meeting the objectives of managing the numerous activities that alter or stress freshwater ecosystems, while at the same time maintaining their ecological functions, is increasingly difficult due to the growing human demand for water. An important point often overlooked in conflicts over permitting land uses is that while there are benefits associated with most industrial, commercial, and recreational activities on public lands, there are also costs associated with these activities. Quantitative information about the costs and benefits of such activities is essential for making informed management decisions. This report focuses on providing tools for examining linkages between surface water, groundwater, and wetland and riparian vegetation. Such tools will enable managers to quantify the costs and benefits of various activities associated with water, land, and river management as well as to examine the physical and biological responses of freshwater ecosystems to factors associated with climate change.

There are 2226 high head dams (greater than 7.6 m high), thousands of smaller dams and detention ponds, tens of thousands of water diversions, and over 90,000 water rights administered on public lands managed by the Forest Service. The Forest Service has the administrative authority to influence the flow regimes and water levels of streams and wetlands on public lands through: (1) the instream flow programs of some states, (2) input (and Section 4e authority) to the hydropower dam relicensing process managed by the Federal Energy Regulatory Commission (FERC 1920), (3) conditioning ditch easements (under the Ditch Bill), (4) restrictions on water extraction and related activities included in land management plans, and (5) terms and conditions applied to such grants of authorization for water diversions, including wells, to minimize damage to scenic and aesthetic values and fish and wildlife habitat and otherwise protect the environment, Federal property, and the public interest (under Federal Land Policy and Management Act).

Although the importance of river flow regime and wetland hydroperiod are well known and a variety of methods have been developed for establishing environmental flows for wetland riverine ecosystems (usually represented by a few species and processes), defining defensible flows and groundwater levels that are necessary to meet management objectives for a site or ecosystem remains challenging. Quantifying how much water is required to maintain desirable characteristics of a wetland, river, or riparian area requires an understanding of the relationships between flow regime and the ecosystem attributes being managed (Richter and others 1997, Richter and others 2003). This often requires an integration of available information about a site and data gathered in the field. Because stream reaches may be gaining (supplied by groundwater) or losing (recharging groundwater), wetlands may be influent or effluent, and water sources may change over time, it is necessary to understand linkages between the factors that influence hydrologic regime and the interactions between surface water, groundwater, and wetland and riparian vegetation. In many cases, this requires examination of available streamflow gauge records, land cover maps or geographic information systems, climatic history, remote sensing imagery, and other information as well as site-specific field measurement of climatic variables, stream discharge, stream and wetland stage, and groundwater levels.

This document will provide guidance to land managers, research scientists, and others tasked with understanding the hydrologic interrelationships between riverine and wetland ecosystems, groundwater, climate, land uses, and stressors. This work begins with a classification and description of major wetland types in the western United States. We then examine factors that influence hydrologic regime in wetlands and rivers, followed by a primer on plant-water relations, plant physiology, and plant and vegetation measurement techniques. Approaches to experimental design and techniques for measuring surface and groundwater are presented along with methods of gathering, processing, and analyzing data from such studies and linking attributes of wetland and riparian vegetation to hydrologic processes. We conclude with several case studies and examples of applications of the tools and methods presented here to systems in the western United States. Though this guide is tailored to the range of wetland and riparian system types in the western United States, the basic principles and methods presented apply to other regions as well.

Wetland and riparian concepts and definitions

Wetlands

While most people are able to identify a forest or grassland as a particular vegetation type, wetlands are more of a challenge to identify, define, and delineate. Wetlands may be dominated by trees, shrubs, herbaceous plants, or mosses or they can be comprised of a mosaic of several of these cover types. Wetland soils range from ancient peat accumulations in high mountain valleys to recent gravel or cobble bars along streams, and they may be highly saline or fresh. Wetland hydrologic regimes are also varied and may include shallow groundwater tables or flooding created by flowing or ponded surface water or groundwater. This range of characteristics makes a simple definition of wetlands elusive. Furthermore, different definitions are appropriate for different purposes (e.g., administrative, functional, academic, or regulatory delineation purposes).

The definition used for Federal regulatory activities of jurisdictional wetlands in the western United States under Section 404 of the Clean Water Act is presented in the U.S. Army Corps of Engineers (COE) Manual (USACE 2009): *wetlands are those areas that are saturated or inundated at a frequency and duration sufficient to support and that under normal circumstances do support a prevalence of vegetation typically adapted for life in saturated soils.* According to this definition, wetlands are saturated or inundated by surface water or groundwater often enough and long enough during the growing season that they support wetland plants. A key tenet of wetland identification using the COE manual is application of the three parameter approach. Jurisdictional wetlands must have positive indicators of each of the three parameters: (1) hydrophytic vegetation, (2) wetland hydrologic regimes, and (3) hydric soils. Hydrophytes—plants with morphological or physiological adaptations for life in saturated soils—are listed in the U.S. Fish and Wildlife Service National Wetlands Inventory "List of Plant Species that Occur in Wetlands" (Reed 1988, Lichvar and Kartesz 2012). Hydric soils are gleyed, or mottled, or may have peat accumulation as described in the COE manual. Hydrologic regimes of wetlands often create flooding, ponding or soil saturation in the upper 30.5 cm of soils for at least two weeks during the growing season, as presented in the COE manual and recent regional supplements (USACE 2009).

A second widely used definition of wetlands was developed by the U.S. Fish and Wildlife Service for the National Wetlands Inventory mapping program (Cowardin and others 1979): *wetlands are lands transitional between terrestrial and aquatic systems where the water table is usually at or near the surface or the land is covered by shallow water. For purposes of this classification wetlands must have one or more of the following three attributes: (1) at least periodically, the land supports predominantly hydrophytes, (2) the substrate is predominantly undrained hydric soil, and (3) the substrate is non-soil and is saturated with water or covered by shallow water at some time during the growing season of the year.* The concept of wetland presented in this definition is broader than the COE definition, as only one of the three parameters need be present for a site to be classed as a wetland. Hydrophytes are plants listed by Reed (1988), hydric soils are defined similarly to the COE, and non-soils are areas that do not support plants because they are too saline, flooded too deeply, or are bare sediment.

Wetland definitions have also been developed by the National Research Council (National Research Council 1995), other Federal agencies such as the Natural Resources Conservation Service, and different regions of the U.S. Forest Service and Bureau of Land Management. All wetland definitions recognize, to one degree or another, the key role of hydrologic processes (e.g., inundation timing, periodicity, and depth) in wetland formation and the resulting suite of distinctive soil and vegetation characteristics.

USDA Forest Service Gen. Tech. Rep. RMRS-GTR-282. 2012

3

For this report, we consider wetlands to be ecosystems that have saturated and anoxic soils for at least two weeks during the growing season over many years.

Riparian

The term riparian has been variously defined and applied for legal, regulatory, and ecological contexts. The term originates from the Latin word *riparius,* which means *of or pertaining to the bank of a river* and is both a geographic concept identifying lands adjacent to streams as well as a hydrologic, geomorphic, and ecological concept identifying sites that are hydrologically and geomorphically influenced by the flowing water of streams. The definition we apply in this assessment closely follows that used by Naiman and Décamps (1997): *it is the portion of the stream channel occurring between the low and high water marks and adjacent terrestrial areas extending from the high water mark toward the uplands where vegetation may be influenced by elevated water tables or flooding.* A key element of this definition is the existence of a physical hydrologic and geomorphic connection, at least intermittently, between the stream and riparian area. In the case of ephemeral streams, this connection may be infrequent and limited to the physical effects of isolated flood events. In contrast, along perennial streams, the stream exerts a more constant and dominant control on ecological function through flooding as well as by influencing water table dynamics. Riparian ecosystems have unique geomorphic characteristics, hydrologic regimes, landforms, biota, and ecological processes that distinguish them from aquatic, isolated wetland, and upland ecosystems.

There are a number of characteristics common to all riparian ecosystems. The first is the periodic or perennial influence of flowing water. Flood events are key drivers of geomorphic, biogeochemical, and biological characteristics of riparian areas (Bowden 1987, Knighton 1998, Pinay and others 2002, Arscott and others 2003) and act to differentiate riparian from other wetland and non-wetland ecosystem types. Riparian areas typically have shallow water tables when compared to adjacent uplands and support distinct vegetation types (Carsey and others 2003), in both perennial and many ephemeral streams (Shaw and Cooper 2008).

Wetland and riparian ecosystem types

Hydrologic regime is the daily, weekly, seasonal, and interannual pattern of flooding, inundation, water table dynamics, and/or soil saturation of a wetland, river, or stream. The hydrologic regime is a function of watershed- and local-scale climate and physical processes that provide water to a site and influence soil conditions and plant water availability. It also includes annual, interannual, and seasonal flooding, ponding or water table depth variation, the ability of water to transport sediment and dissolved materials and nutrients on a seasonal and long-term basis, and the rate of change in transitioning from seasonal high to low water levels. The hydrologic regime of a wetland influences the wetland type and its ecological functioning more than other factors. For example, it influences ecosystem productivity, decomposition of organic matter, mineral sediment erosion and deposition, seasonality of drought, depth of flooding (Pinay and Naiman 1991, Mitsch and Gosselink 2000, Weltzin and others 2000), rates of denitrification (Pinay and others 2007), and site biotic composition (Cooper and Andrus 1994, Seabloom and others 1998). At local scales, the hydrologic regime operates as a driver of ecological structure and function, facilitating peat accumulation or mineral sediment deposition to create landforms, while at larger scales it shapes landscape-scale patterns of wetland and riparian ecosystem abundance and distribution. Collectively, climate, geology, and hydrologic regime structure the template from which wetlands and riparian ecosystems form, develop, and function.

There are five main inland wetland and riparian ecosystem types in the western United States, and they vary in their relative abundance, distribution on the landscape, vegetation composition, structure and dynamics, and functional characteristics. These types are: (1) fens, (2) wet meadows, (3) marshes, (4) salt flats, and (5) riparian areas. Each type can also be subdivided based upon hydrologic regime, vegetation, geochemistry, and other factors into a number of different subtypes, typically defined by dominant plants, plant associations, or communities. Although frequently managed and regulated as a single resource, wetland and riparian ecosystem types differ widely in their hydrologic regimes and processes, vegetation, and functional characteristics and in their responses to stresses posed by humans and natural disturbances. The processes occurring within a wetland or river are typically not controlled, thus they cannot often be managed, at the scale of the site. Conditions and activities that influence water and sediment within the contributing groundwater and surface watershed must be considered.

Fens

Fens have perennial groundwater inflows that maintain water tables at or near the ground surface. This constant saturation retards the decomposition of organic matter and allows for peat accumulation. Being groundwater driven, fens vary considerably in the chemical content and pH of the source water. The chemical composition of groundwater is controlled by the mineral composition of bedrock and unconsolidated sediments in the contributing watershed. Fens form in a variety of landscape settings and are among the most floristically diverse ecosystems in the region, and many support rare and widely disjunct species (Cooper and Andrus 1994, Cooper 1996, Cooper and Sanderson 1997, Chadde and others 1998, Weber 2001, Bedford and Godwin 2003, Hiedel and Laursen 2003). In contrast to riparian areas, little mineral sediment generally moves into or out of fens, and they are geomorphically stable on a time scale of millennia. Fens may occur within or adjacent to riparian areas or other wetland types in valley bottoms, but they have independent groundwater sources and are not dependent directly upon stream water. Fens are typically dominated by herbaceous monocots in the family Cyperaceae (sedge family) and may have a continuous carpet of mosses. However, some fens have a canopy of shrubs or conifer trees. Fens have formed in a variety of landforms, including slopes and basins, and create landforms such as floating mats, hummocks, strings, and pools through the process of peat accumulation. Fens provide critical habitat for amphibians, many small mammals, aquatic invertebrates, and passerine birds (Figure 1-1).

Most fens have formed in locations where groundwater discharges to the surface. These are the most stable spring complexes in the region, and plants colonize these springs, blanketing the water source with layers of roots, rhizomes, leaf and stem litter, and other undecomposed plant remains that form the peat bodies. The water flows vertically up through the peat body and laterally through layers within the peat. Some peatlands have sheet flow across their surfaces in early through mid-summer, others have distinctive water-tracks, while others rarely have surface water. Fens that formed in basins may have pools or ponds that are being encroached upon by floating mats that will eventually fill the basin with peat.

Figure 1-1. Fens. (A) is East Lost Park in the Taryall Mountains, Colorado. (B) is Green Mountain Trail pond fen with a floating mat in Rocky Mountain National Park, Colorado. (C) is High Creek fen, an extremely rich fen in South Park, Colorado. (D) is Drosera fen in Yosemite National Park, California. (E) is Long Meadow in Rocky Mountain National Park, Colorado.

Wet meadows

Wet meadows are widespread and likely cover more area in the mountainous western United States than all other wetland types combined. They occur from alpine areas to plains and from intermountain parks and basins to foothills. Despite their relative abundance, few studies have examined their hydrologic and edaphic characteristics or vegetation dynamics. Although wet meadows typically have seasonally saturated soils, they lack the perennial high water tables of fens or the large seasonal and inter-annual water table fluctuations of marshes, and they do not form peat soils. Many natural wet meadows are managed for livestock forage and hay production, and in agricultural areas, wet meadows have been created by application of water from ditches. Most wet meadows are dominated by herbaceous plants, particularly *Juncus arcticus*, *Carex nebraskensis*, and *C. lanuginosa*. Woody plants, such as *Pentaphylloides (Potentilla) floribunda* (cinquefoil) and a number of *Salix* (willow) species, may also be present (Figures 1-2A through D).

Figure 1-2. Wet meadows. (A) is Convict Creek valley, Sierra Nevada, California. (B) is Log Meadow in Sequoia National Park, California. (C) is in the San Luis Valley, Colorado. (D) is a meadow at the foot of Great Basin range, Nevada.

USDA Forest Service Gen. Tech. Rep. RMRS-GTR-282. 2012

7

Marshes

Marshes form in depressions and include such diversely named regional wetland types as prairie potholes, playas, vernal pools, lacustrine fringes, and oxbow lakes on river flood plains (Figure 1-3). Marsh hydrologic regimes are variable, with both prolonged periods of inundation and desiccation (Winter and Rosenberry 1998, Winter and others 2001). Marshes periodically have deep standing water (>1m), which limits the species that occur to aquatic and wetland species tolerant of submersion or deep inundation. Because some marshes are terminal basins with surface water inflows but little or no outflow, their chemical content varies from freshwater to saline and this influences plant and animal species composition, litter decomposition rates, and productivity of plants, aquatic invertebrates, and larger animals (Thormann and others 1999). Hydrologic variability, water depth, and salinity are key factors determining the species composition of marshes, both spatially within and among marsh complexes and temporally from wet to dry climate periods (van der Valk and others 1994, Seabloom and others 1998, Smith and Haukos 2002). Seed banks play a particularly important role in marsh vegetation dynamics (van der Valk and Davis 1976, Smith and Kadlec 1983, Wilson and others 1993), with large fluctuations in species composition commonly occurring over relatively short time scales. The large water depth gradients also generate distinct vegetation zonation patterns in many marshes (Johnson and others 1987, Lenssen and others 1999), and their diverse hydrologic regimes and vegetation types provide critical habitat for waterfowl, shorebirds, and amphibians.

Figure 1-3. Marshes. (A) is bulrush-dominated marsh with prominent zonation near Denver, Colorado. (B) is large marsh with fringing vegetation and floating bulrush clones in North Dakota. (C) is marsh on Yellowstone National Park's northern range, Wyoming. (D) is Heart Lake marsh in Yellowstone National Park, Wyoming.

Salt flats

Salt flats are widespread at low elevations through the West, particularly in intermountain basins (Figure 1-4). They form where high soil salt concentrations occur near the soil surface. The two main processes that promote salt flat formation are surface water evaporation from basins with fine-textured soils, and evaporation from the capillary fringe of a shallow water table that leads to salt accumulation on the soil surface. The combination of high salt concentrations and saturated soils creates difficult growing conditions for plants; consequently, plant cover and productivity are typically low, and vegetation composition is limited to species tolerant of both high salt concentrations and saturated soils (Dodd and Coupland 1966, Ungar 1966, Ungar 1974). Salt flats may also occur in saline soils on river floodplains, particularly along rivers that rarely experience overbank flooding due to upstream dams or diversions (Jolly and others 1993). Ironically, because salt flats are marginal for forage production and are unsuitable for crops, they have been spared many of the anthropogenic impacts affecting other wetland types, other than dewatering. Salt flats are typically dominated by grasses such as *Distichlis spicata* (salt grass), *Sporobolus airoides* (alkali sacaton), *Spartina gracilis* (cordgrass), and herbaceous dicots such as *Triglochin maritimum* (arrow grass) and other halophytes, many of which also occur in coastal salt marshes. When flooded, salt flats support high densities of aquatic invertebrates and are important habitat for migratory waterfowl and shorebirds.

Riparian areas

Riparian ecosystems have diverse landforms, stream sizes, valley gradients, hydrologic regimes, vegetation, and ecological functions. They vary from low-gradient water tracks running adjacent to peatlands at high elevations; to steep-gradient, small-order headwater mountain streams; to ephemeral streams in mountain foothills; to those along broad, low-gradient alluvial rivers in the Great Plains (National Research Council 2002).

Figure 1-4. Salt flats. (A) is in the San Luis Valley, Colorado. (B) is in North Park, Colorado.

The principal characteristic unifying riparian ecosystems is the presence of moving water, which has the potential to erode, transport, and deposit sediment and to create distinctive landforms such as point bars, floodplains, and abandoned channels. These landforms and the fluvial processes that influence their creation, destruction, and turnover are critical for the establishment and persistence of riparian plants.

The energy of flowing water is a key variable influencing riparian structure and function. The frequency, magnitude, and energy of floods, which vary widely due to differences in basin size, topography, and climatic regime, affect all ecological processes from nutrient cycling to plant establishment to rates of channel migration, floodplain development, and riparian forest formation and turnover (Karrenberg and others 2002, Cooper and others 2003a, Adair and others 2004).

The hydrologic regimes of streams also vary widely and provide important constraints and opportunities for riparian and aquatic organisms. Though all streams receive multiple sources of water over time, streams may be broadly divided into classes based upon the dominant sources of water that influence the hydrologic regime (Poff 1996). Hydrologic regimes of riparian areas vary from relatively stable, groundwater-driven flows, such as those in the Nebraska Sandhills (Bio/West 1986, Winter 1999), to infrequent and unpredictable flash floods associated with ephemeral streams located throughout the mountainous and semiarid western United States (Friedman and Lee 2002, Shaw and Cooper 2008).

Seasonal flooding may be governed by frontal weather systems, monsoons, convective storms, or seasonal changes in temperature that govern snowmelt regimes (Wohl 2000). An understanding of the sources of flow to a stream, the magnitude and frequency of those flows, and the seasonal and interannual timing in those flows is important in understanding principal forces that govern species composition, turnover, and ecological functioning of riparian areas and their potential responses to changing stream flow or groundwater regimes.

Many riparian areas are dominated by woody plants, in particular by species of *Populus* (cottonwood) at the lowest elevation and *Salix* species at all elevations. At mid elevations, species of *Alnus* (alder) and *Betula fontinalis* (birch) may be common. The smallest springs and brooks may be dominated by herbaceous plants, including species of *Mertensia* (blue bells), *Glyceria* (mannagrass), *Senecio* (groundsel), and *Mimulus* (monkey flower). In more arid regions such as the Sonoran desert, riparian areas may be characterized as forests dominated by species of *Juglans* (walnut), *Platanus* (sycamore), *Populus*, *Salix*, *Prosopis* (mesquite) and *Fraxinus* (ash) (Figure 1-5).

Vegetation of wetland and riparian areas in the western United States

Wetlands and riparian areas support a variety of plant species and community types found nowhere else in the West. For example, 183 (31%) of the nearly 600 ecological system types (groups of plant community types that co-occur in similar ecological settings) defined by Comer and others (2003) in their analysis of Rocky Mountain region vegetation are wetlands, even though wetlands occupy only 1 to 2 percent of the western landscape.

In order to understand the range of wetland forms throughout the western United States, we compiled vegetation data from 5266 wetland and riparian plots from the Rocky Mountain region (including the Great Plains). To understand the structure and variety of wetlands represented in this large data set and to see relative relationships between wetland types, we performed an indirect ordination on the riparian and wetlands plots (Figure 1-6) using detrended correspondence analysis (DCA; McCune and Mefford 1999). Plot data include species composition and percent canopy coverage by species.

Figure 1-5. Riparian areas. (A) is small, glaciated stream valley on east side of Wind River Range, Wyoming. (B) is subalpine stream in Rocky Mountain National Park, Colorado. (C) is the Blue River downstream from Dillon, Colorado. (D) is Yampa River in Deer Lodge Park, Colorado. (E) is Lithodendron Creek in Petrified Forest National Park, Arizona.

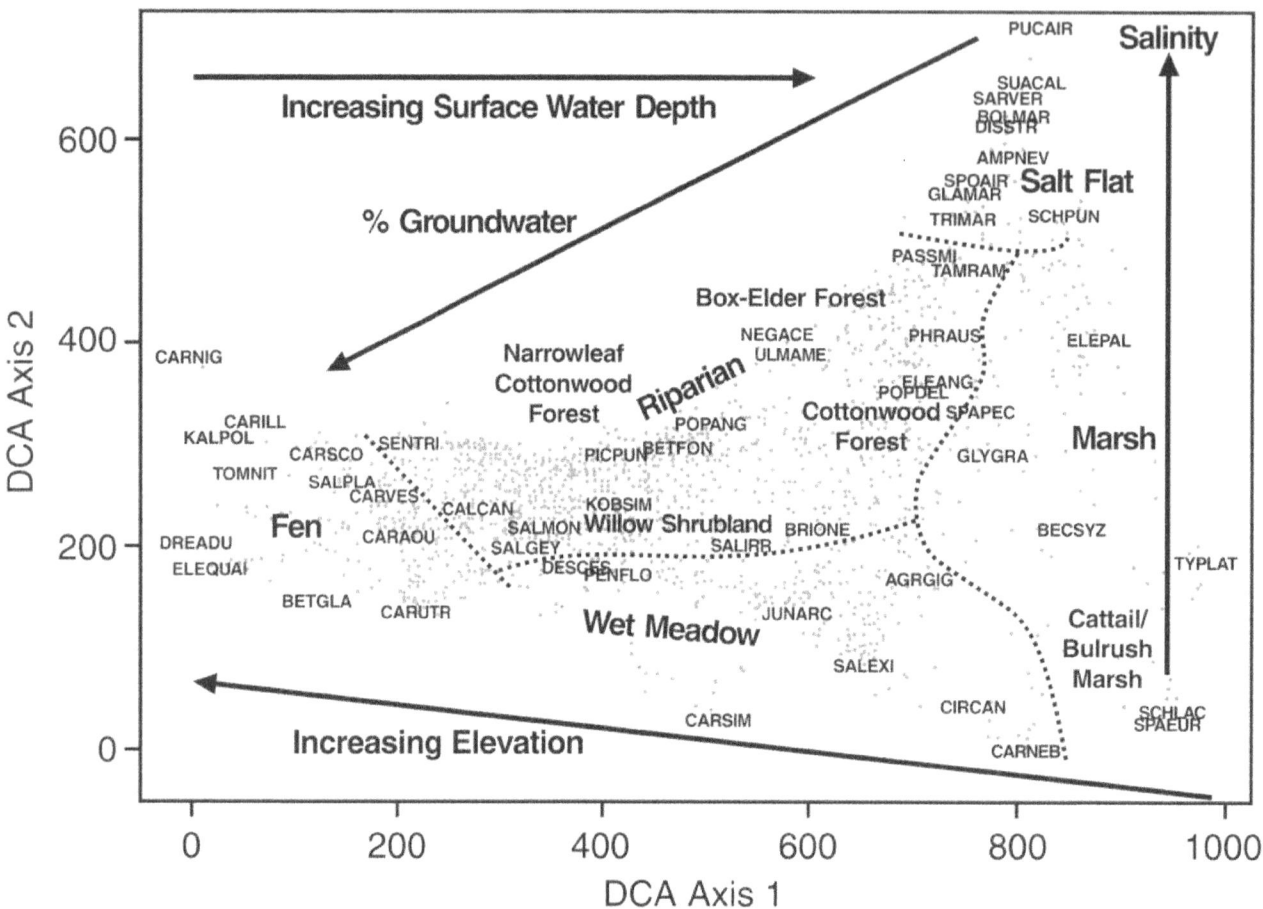

Figure 1-6. Detrended Correspondence Analysis (DCA) ordination of wetland and riparian communities. The location of major wetland communities are identified in the graph. The main direction of environmental variation is shown using arrows. Centroids of diagnostic and common plant species are shown using the following abbreviations: CARNIG = *Carex nigricans*, DREADU = *Drepanocladus aduncus*, ELEQUI = *Eleocharis quinqueflora*, KALPOL = *Kalmia polifolia*, CARILL = *Carex illota*, TOMNIT = *Tomenthypnum nitens*, BETFON = *Betula fontinalis*, CARSCO = *Carex scopulorum*, SALPLA = *Salix planifolia*, CARVES = *Carex vesicaria*, SENTRI = *Senecio triangularis*, CARAQU = *Carex aquatilis*, CARUTR = *Carex utriculata*, CALCAN = *Calamagrostis canadensis*, SALMON = *Salix monticola*, SALGEY = *Salix monticola*, DESCES = *Deschampsia cespitosa*, PENFLO = *Pentaphylloides floribunda*, PICPUN = *Picea pungens*, BETGLA = *Betula glandulosa*, POPANG = *Populus angustifolia*, SALIRR = *Salix irrorata*, CARSIM = *Carex simulata*, JUNARC = *Juncus arcticus*, SALEXI = *Salix exigua*, CARNEB = *Carex nebraskensis*, CIRCAN = *Cirsium canadensis*, AGRGIG = *Agrostis gigantea*, BROINE = *Bromopsis inermis*, NEGACE = *Negundo aceroides*, ULMAME = *Ulmus americanus*, POPDEL = *Populus deltoides*, ELEANG = *Eleagnus angustifolia*, PHRAUS = *Phragmites australis*, SPAEUR = TAMRAM = *Tamarix* spp., PASSMI = *Pascopyrum smithii*, PUCAIR = *Puccinellia airoides*, SUACAL = *Sueda calcioliformia*, SARVER = *Sarcobatus vermiculatus*, BOLMAR = *Bolboschoenus maritimus*, DISSTR = *Distichlis stricta*, AMPNEV = *Amphiscirpus nevadensis*, SPOAIR = *Sporobolus airoides*, GLAMAR = *Glaux maritimus*, TRIMAR = *Triglochin maritimum*, SCHPUN = *Schoenoplectus pungens*, ELEPAL = *Eleocharis palustris*, SPAPEC = *Spartina pectinatus*, GLYGRA = *Glyceria grandis*, BECSYZ = *Beckmannia syzygachne*, TYPLAT = *Typha latifolia*, SCHLAC = *Schoenoplectus lacustris*, SPAEUR = *Sparganium eurycarpum*.

DCA places observations (vegetation plots) in a multi-dimensional ordination space, with plots with the most similar floristic composition occurring close to each other, and those most dissimilar being farthest apart. The axes are in standard deviation units, with 200 to 400 units indicating a complete species turnover; for example, a plot at 100 on axis 1 (x-axis) and a plot at 500 on axis 1 would likely have no species in common.

Variation in species composition along axis 1 represents an increase in elevation and an increase in water table permanence (from right to left). High-elevation sites with perennially high water tables occur on the left, and low-elevation sites with varying water tables occur on the right. Variation in axis 2 appears to be driven by water chemistry, with saline sites near the top and freshwater sites toward the bottom. Each plot is represented by one point, and the centers of abundance of key indicator plant species are shown in the ordination space.

The five major wetland types break out distinctly in the ordination space. Fens are on the far left, salt flats are on the top right, marshes are on the bottom right, wet meadows are on the bottom center, and riparian areas are along a band that extends from the top right to the left. The riparian continuum includes springs and small headwater mountain streams with herbaceous communities dominated by species such as *Senecio triangularis*, *Salix geyeriana*, and other *Salix*-dominated thickets (also called carrs) along small- to medium-sized mountain rivers, *Populus angustifolia* (narrowleaf cottonwood)- and *Picea pungens* (blue spruce)-dominated forests along mid-elevation mountain rivers, *Acer negundo* (box elder) forests in canyons, and *Salix irrorata* and *Populus deltoides* forests with *Tamarix* (salt cedar) and *Elaeagnus angustifolia* (Russian olive) along foothills and plains streams.

A range of fen types are separated in the ordination, including basin fens dominated by sedges such as *Carex utriculata* and *Carex vesicaria*; sloping fens dominated by *Eleocharis quinqueflora* (spike rush), *Carex aquatilis*, and *Carex illota*; and wooded fens with *Kalmia polifolia* (swamp laurel) and *Salix planifolia*. There are a range of wet meadows dominated by *Deschampsia cespitosa* (tufted hairgrass) at high elevation, *Juncus arcticus* (arctic rush) at mid to low elevation, and *Carex nebraskensis* at the lowest elevations. A wide range of marshes occur, and due to the deep water, many are monocultures or have very low floristic diversity, creating a wide spread in the plots. The deepest water sites have *Schoenoplectus lacustris* (bulrush) and *Typha latifolia* (cattail), while more shallow water sites are dominated by *Eleocharis palustris* and *Schoenoplectus pungens* (three-square). The latter species can also occupy marshes that are highly saline and plot adjacent to salt flats. The salt flat communities are also species poor, and many monocultures occur.

Chapter 2: Hydrologic Regime and Factors That Govern Hydrologic Processes

The principal source of water supplying most streams, groundwater, and wetlands is precipitation in the form of rain and snow. Five main hydrologic processes influence all riparian and wetland ecosystems: (1) the amount, timing, and type of precipitation, (2) groundwater recharge, (3) groundwater discharge, (4) surface water runoff, including stream flow, and (5) evapotranspiration (Figure 2-1). Regions with higher total annual precipitation typically have higher annual stream flows, a higher proportion of perennial streams, and, in many areas, perennial groundwater flow systems (Lins 1997).

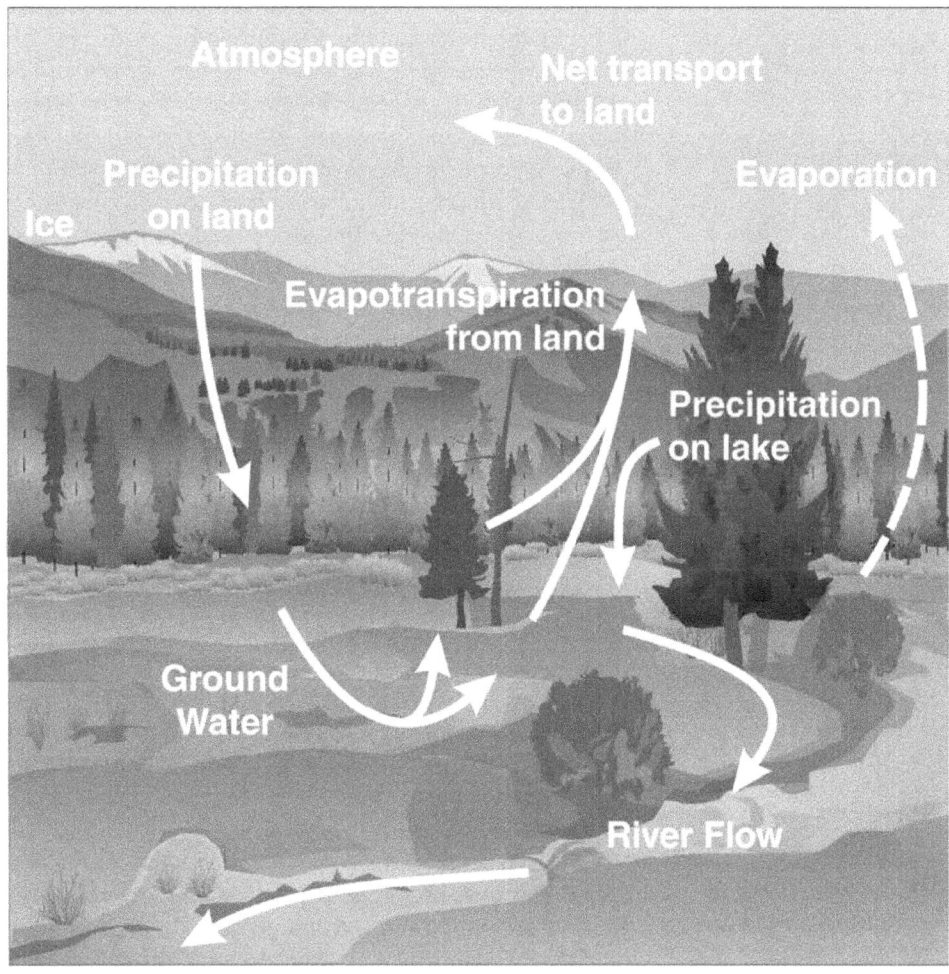

Figure 2-1. Components of the hydrologic cycle that influence riparian and wetland ecosystems, including precipitation, runoff and river flow, groundwater flow, and evaporation and evapotranspiration.

Surface water runoff occurs whenever the rate of water delivery to soils—from precipitation, snowmelt, or groundwater discharge—exceeds the rate and capacity of the soil for infiltration. Infiltration rates are influenced by soil texture and structure, depth to bedrock, vegetation, and soil bioturbation. Shallow bedrock or permafrost forms a confining or impermeable layer that slows or detains groundwater, which leads to pooling. Coarse textured soils and those with high porosity and/or abundant macropores generally have higher infiltration rates and higher hydraulic conductivity. Vegetation can slow the flow of surface water over the land surface and facilitate infiltration. Groundwater recharge is the process of water infiltrating through the soil profile to a local or regional water table. The water table is the top of the most shallow groundwater flow system. Below the water table, soils are saturated; above the water table, the soils are unsaturated and under tension. Groundwater is present beneath most landscapes where there is shallow bedrock or thick deposits of unconsolidated material such as moraine, alluvium, or colluvium underlain by an impermeable layer.

Groundwater flows along pressure gradients and discharges to the ground surface where the water table intersects the soil surface. Groundwater discharge may be caused by physical features such as a decrease in depth to bedrock, which forces groundwater to the surface, or a reduction in land slope that reduces the rate of groundwater flow causing the water table to rise. Groundwater discharge may occur at a point, which is typically termed a spring, but it may also occur along broad seepage faces (e.g., along a valley wall) or beneath streams, lakes, or fens. Surface water and groundwater are subject to evaporation as well as transpiration by plants, and the combined loss of water to the atmosphere is termed evapotranspiration (ET).

The distribution, abundance, and types of wetland and riparian ecosystems that occur within a watershed are linked to and dependent upon water availability. Precipitation varies dramatically across the western United States, with the highest annual precipitation totals occurring west of the Cascade Range and Sierra Nevada and in some high mountain regions occurring in the Great Basin and Rocky Mountains (Figure 2-2A). Annual precipitation totals also vary markedly along elevation gradients, even over short distances due to orographic lift, producing higher precipitation totals on mountains, with rain shadows forming in the leeward side of mountains. High mountain watersheds may receive five times more precipitation than valleys and basins, as illustrated for North Park and the Park Range in northern Colorado (Figure 2-2B). Winter storms typically move from the Pacific Ocean to the east, and west-facing mountain slopes receive higher precipitation totals than eastern slopes, as occurs in the Sierra Nevada and Cascades.

The seasonality of precipitation also varies across the region. The far west and northern Rocky Mountains receive mainly winter precipitation. On the Great Plains, the precipitation peak is in spring and early summer from air masses moving north from the Gulf of Mexico (Figure 2-2C). In the Southwest, including the southern Rocky Mountains, there is a winter peak and a second late summer monsoon precipitation (rain) peak. The seasonality of precipitation, along with the temperature-controlled timing of snowmelt, influences the pattern of stream flow, groundwater recharge and discharge, and surface runoff that fills basins.

A deep snowpack may accumulate in high mountain regions from October through April and melt in spring to produce a snowmelt-driven flow peak and an important period of groundwater recharge (Figure 2-3). Once the watershed snowpack has melted, streamflow is supported by groundwater discharge and runoff from precipitation events; lower stream flows occur during most of the summer. The majority of snowmelt-driven streams are relatively small and occur at high elevations, however, they feed the relatively few perennial regional streams. Some perennial rivers in the southwestern United States have high flow events due to both snowmelt and monsoon rains, as illustrated by the San Miguel River in southwestern Colorado (Figure 2-3).

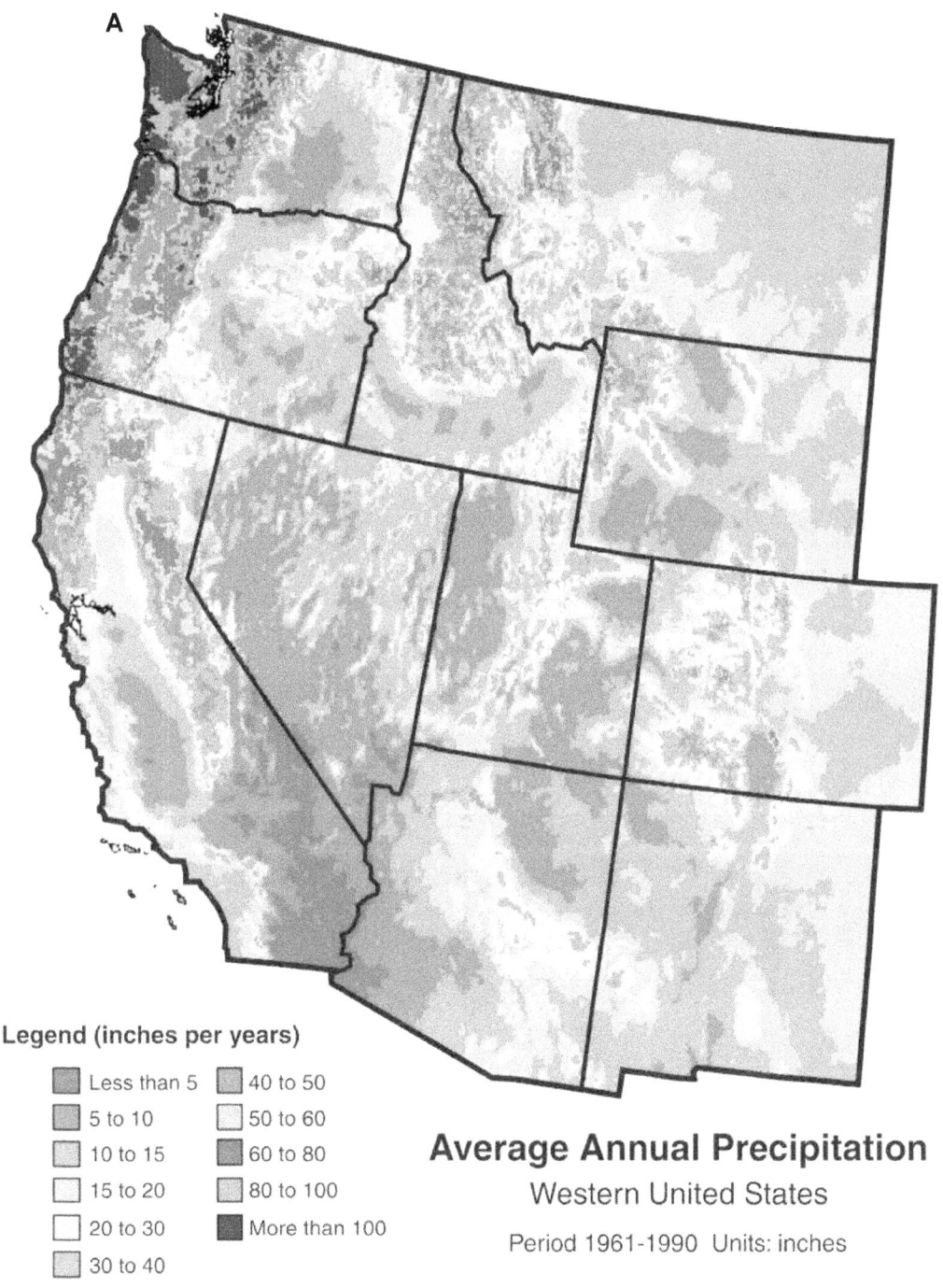

A

Legend (inches per years)

- Less than 5
- 5 to 10
- 10 to 15
- 15 to 20
- 20 to 30
- 30 to 40
- 40 to 50
- 50 to 60
- 60 to 80
- 80 to 100
- More than 100

Average Annual Precipitation
Western United States
Period 1961-1990 Units: inches

Figure 2-2A. Average annual precipitation for the western United States (http://www.wrcc.dri.edu/pcpn/westus_precip.gif).

USDA Forest Service Gen. Tech. Rep. RMRS-GTR-282. 2012

17

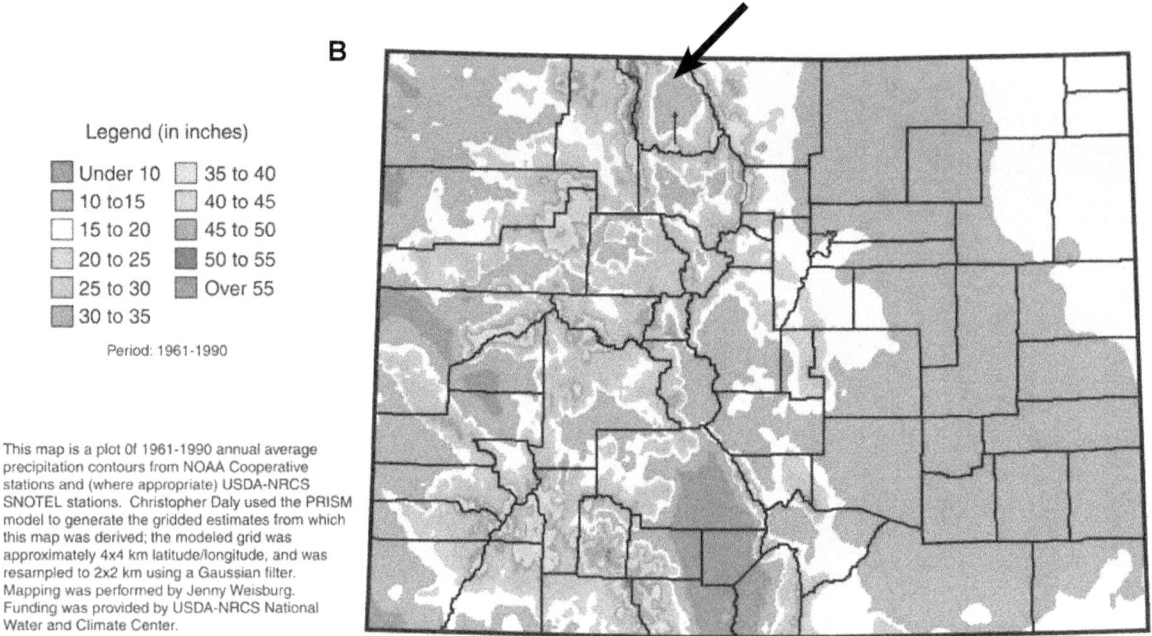

B

Legend (in inches)

- Under 10
- 10 to15
- 15 to 20
- 20 to 25
- 25 to 30
- 30 to 35
- 35 to 40
- 40 to 45
- 45 to 50
- 50 to 55
- Over 55

Period: 1961-1990

This map is a plot of 1961-1990 annual average precipitation contours from NOAA Cooperative stations and (where appropriate) USDA-NRCS SNOTEL stations. Christopher Daly used the PRISM model to generate the gridded estimates from which this map was derived; the modeled grid was approximately 4x4 km latitude/longitude, and was resampled to 2x2 km using a Gaussian filter. Mapping was performed by Jenny Weisburg. Funding was provided by USDA-NRCS National Water and Climate Center.

Figure 2-2B. Average annual precipitation for Colorado. Arrow points to North Park, a low-elevation basin surrounded by mountains. The Park Range on the west side of North Park receives approximately five times the average annual precipitation as the Park floor.

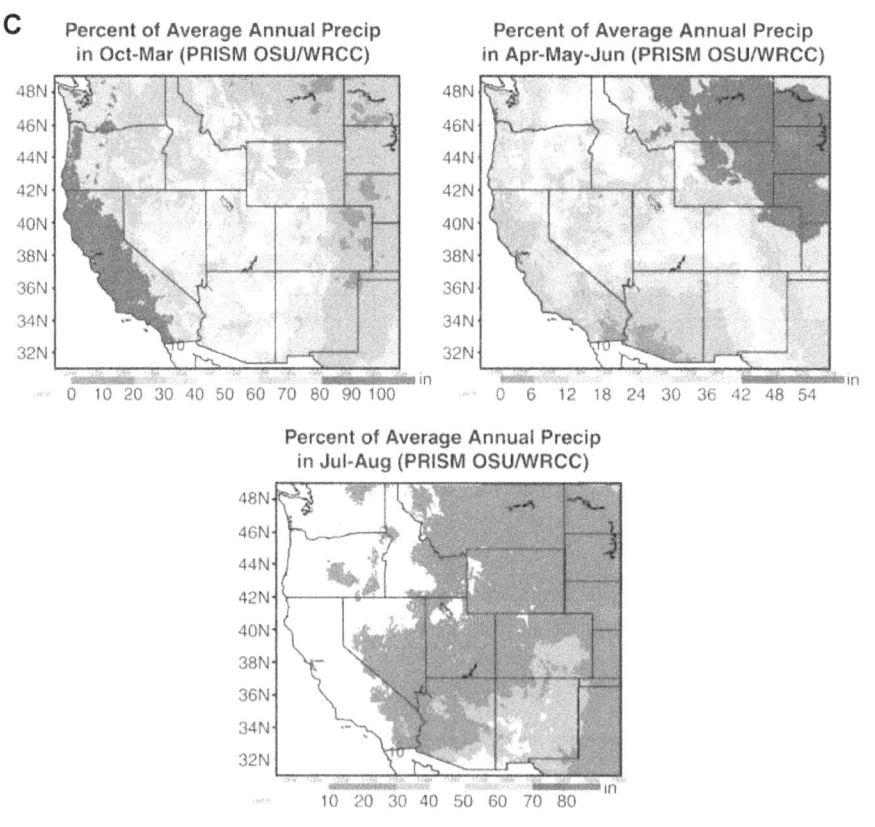

Figure 2-2C. Percentage of winter (top left), spring (top right), and late summer (bottom) precipitation for the western United States.

USDA Forest Service Gen. Tech. Rep. RMRS-GTR-282. 2012

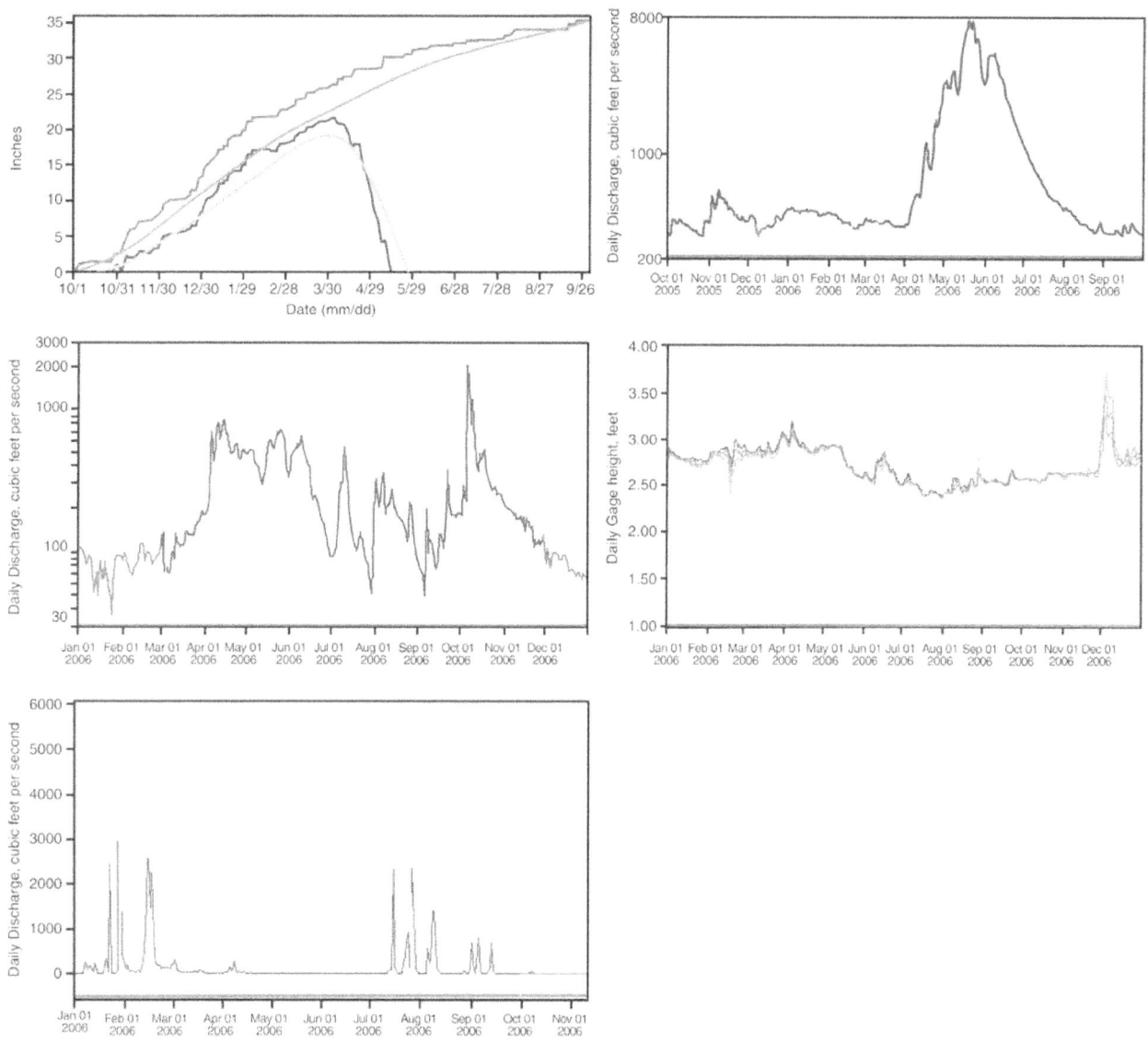

Figure 2-3. (Top left) Snow water equivalent (SWE) (blue line) in inches at the Snake River station, Wyoming, for 2006. Red line is total cumulative precipitation in 2006, orange is mean cumulative precipitation for 1971 to 2000, and grey is mean cumulative SWE for 1971 to 2000. (Top right) Mean daily discharge of the Snake River at the Flagg Ranch, Wyoming, showing the timing of stream peak flow relative to snowpack at the Snake River snow course. (Center left) Mean daily flow for the San Miguel River, Uruvan, Colorado, showing an early summer snowmelt runoff peak with additional monsoon rain peaks in August, September, and October. (Center right) Niobrara River, Sparks, Nebraska, showing that river stage height for this spring-fed river varies little. Bottom left: Puerco River at Chambers, Arizona, is an ephemeral stream flowing only following snowmelt, or rain events. In 2008, there were three large snowmelt-driven floods in January through February and six monsoon rain-driven floods in late summer.

USDA Forest Service Gen. Tech. Rep. RMRS-GTR-282. 2012

19

Ephemeral streams flow only following runoff events, triggered by either snowmelt or summer rainstorms or both, as illustrated by the Puerco River in Arizona (Figure 2-3). Other streams may be spring-fed (groundwater supported) and have little annual variation, as illustrated by the Niobrara River in Nebraska (Figure 2-4).

Most streams are connected to and interact with groundwater systems. Most perennial streams are fed by surface runoff as well as groundwater, the latter of which sustains flow during periods lacking precipitation or snowmelt. Streams may gain groundwater through their banks and from under their bed, and are termed gaining streams, as their flow increases as it moves downstream due to groundwater inputs (Figure 2-5: top panel). Streams may also lose water from their bed and banks to the groundwater system, and are termed losing streams, as their surface flow decreases in a downstream direction (Figure 2-5: middle panel). Many losing streams are intermittent or ephemeral in at least some reaches. Many streams are gaining in some reaches and losing in other reaches, supporting very different environments in the stream and on the floodplains. Some ephemeral streams in arid regions with deep water tables, or those on bedrock, are not connected to the water table (Figure 2-5: bottom panel). The relationship of a stream and the local or regional water table will influence stream and riparian zone functions.

Valley form (width, slope, and depth to bedrock) may influence whether a stream is gaining or losing along its length. As valleys narrow in a downstream direction, groundwater may upwell, causing the stream to gain groundwater and flow on the surface or increase in volume (Figure 2-6). As valleys widen, surface water may be lost to the deeper, wider alluvium, and the stream may decrease in volume or disappear completely as water infiltrates into the groundwater. This process of valley-controlled gaining and losing may result in secondary biogeochemical processes that can influence ecological patterns on the landscape, such as the presence of particular riparian communities (Harner and Stanford 2003).

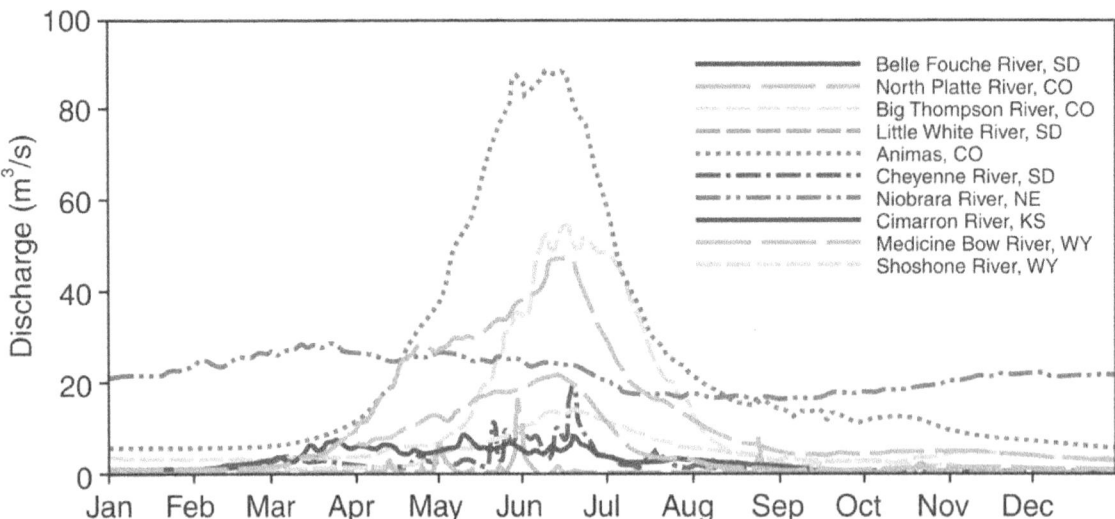

Figure 2-4. Typical discharge patterns for a range of rivers. Rivers originating in the high mountains such as the Animas and Shoshone have large, snowmelt-dominated peak discharges. In contrast, flood flows along Great Plains rivers such as the Cimarron and Cheyenne are driven principally by rain events and are frequently less predictable and more "flashy" in nature. The Niobrara River and other rivers in the Sandhill region of Nebraska have continuous groundwater inputs with relatively constant stream discharge.

Gaining Stream

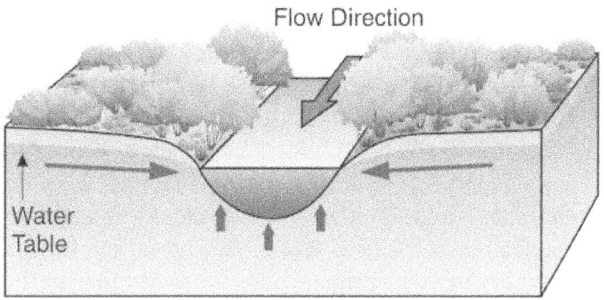

Losing Stream

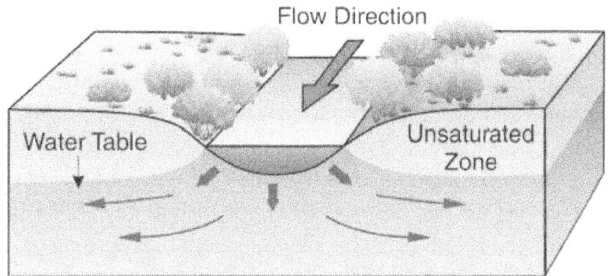

Disconnected Stream

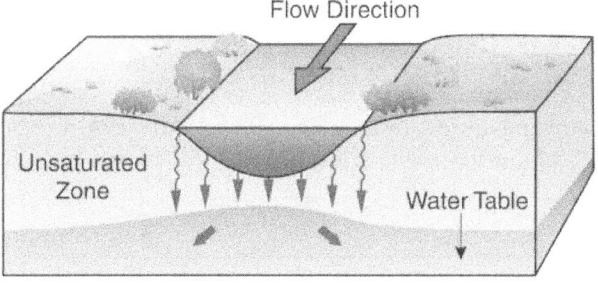

Figure 2-5. Connections of streams and groundwater (modified from Winter and others 1996).

Figure 2-6. Valley form may influence whether a reach is gaining or losing. Where valleys narrow, groundwater may upwell and result in gaining stream reaches; where valleys widen, streams may lose water to deeper and wider alluvium. This pattern may be revealed by higher abundance and productivity of riparian vegetation in upwelling valley segments. From Harner and Sanford (2003).

USDA Forest Service Gen. Tech. Rep. RMRS-GTR-282. 2012

21

Riparian vegetation along gaining and losing stream reaches may respond differently to altered flow regimes. Small changes in streamflow may have significant effects on riparian vegetation along a losing reach, whereas gaining reaches may have limited effects from certain altered flow regimes.

Wetlands form where surface water ponds or where groundwater saturates soils. In the prairie pothole region of North Dakota, rain and snowmelt runoff fills basins (Figure 2-7) where groundwater recharge occurs. Groundwater flow paths may connect basin A through basin B to basin C (Figure 2-7). Basin A is likely to be intermittently flooded, while the long groundwater flow path may produce perennial saturation or inundation in basins B and C. Because basin B has water flow through, it does not accumulate salts; then again, basin C is a terminal basin with water leaving by evaporation, and high salt concentrations may occur. The complex recharge, groundwater flow, and discharge patterns produce a range of wetland types. Recharge basins (A) support fresh water marshes, flow through basins (B) may be peat accumulating fens, and discharge areas (C) may support salt marshes. The range of wetland types and the chemistry of surface water and groundwater will vary depending upon the bedrock or surficial material that water flows through and the availability and the availability and solubility of salts and other materials.

Four wetland hydrologic regimes can be conceptualized: (1) groundwater depression wetland, (2) groundwater slope wetland, (3) surface water depression wetland, and (4) surface water slope wetland (Novitzki 1982: Figure 2-8). Groundwater depression wetlands form in basins that intercept the regional or local water table. If the groundwater flow is seasonal or if water levels vary, the wetland likely will be a marsh (Figure 2-8: top left panel; Figure 2-9: top panel). If the flow is perennial, without considerable annual and interannual variation, a fen will likely form. Groundwater slope wetlands form where groundwater discharges at a geologic discontinuity or toe slope, and wet meadows or fens are formed (Figure 2-8: top right). Where surface flows occur in channels, riparian zones may develop (Figure 2-8: bottom right). Surface water slopes can also form on the fringes of lakes (Figure 2-9). Surface water depressions are formed by overland flow or as terminal sumps for streams and would form marshes or salt flats (Figure 2-8: bottom left panel).

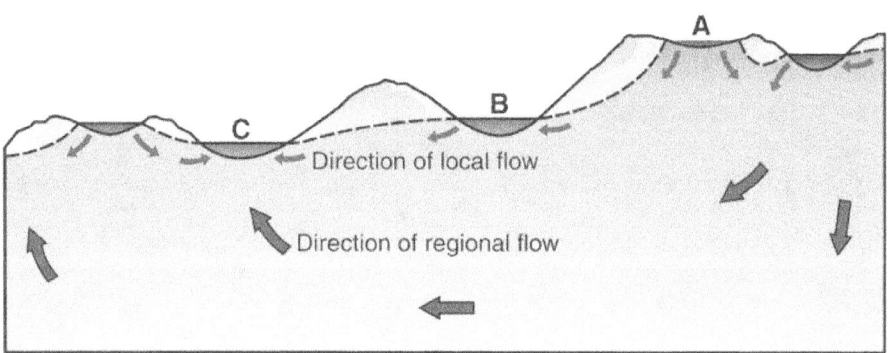

Figure 2-7. Surface water and groundwater flow in complex, glaciated terrain. Precipitation runoff fills ponds (A) that recharge groundwater that flows through basin B and discharges into basin C. Modified from Winter (1989).

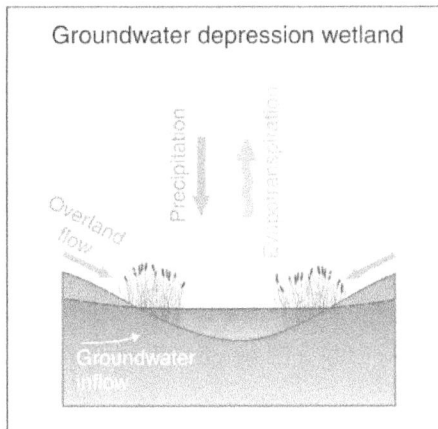
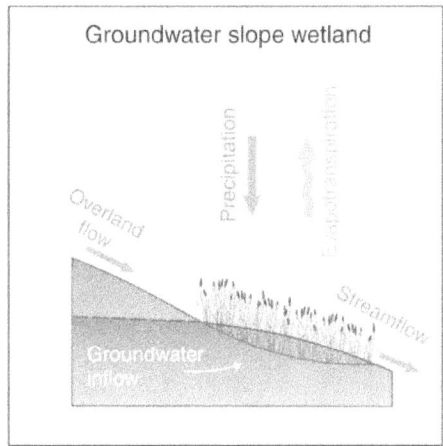
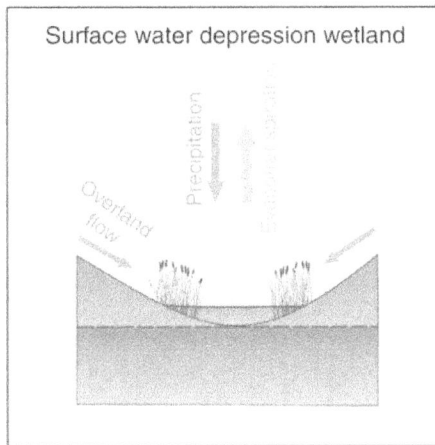
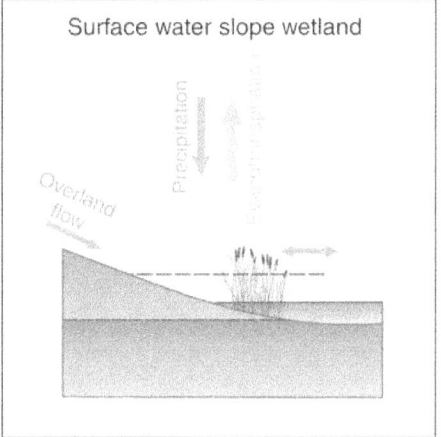

Figure 2-8. Hydrologic processes supporting four main wetland types. Modified from Novitzki (1978).

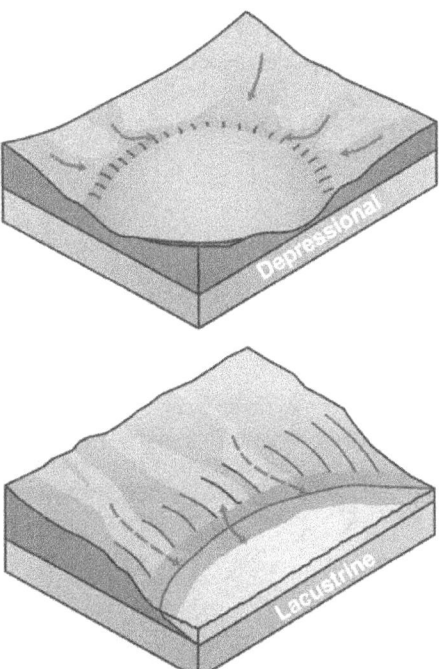

Figure 2-9. Geomorphic settings for marsh development. Arrows indicate principal water fluxes of water. Modified from Brinson and Malvarez (2002).

USDA Forest Service Gen. Tech. Rep. RMRS-GTR-282. 2012

23

Salt flat wetlands may form through several hydrologic processes. Salt accumulates in wetlands associated with seasonal ponds that occur on low-permeability soils or seasonally flooded lake margins (Figure 2-10A). Salt also accumulates where groundwater discharges to the soil surface (Figures 2-10B and C) or is elevated by capillary rise and evaporates, or it occurs where the capillary fringe reaches the surface and water evaporates (2-10D). High salt accumulation can only occur where flow does not remove salts from a site, and in such places, salt may accumulate even when source water has low ion concentrations (Jolly and others 1993).

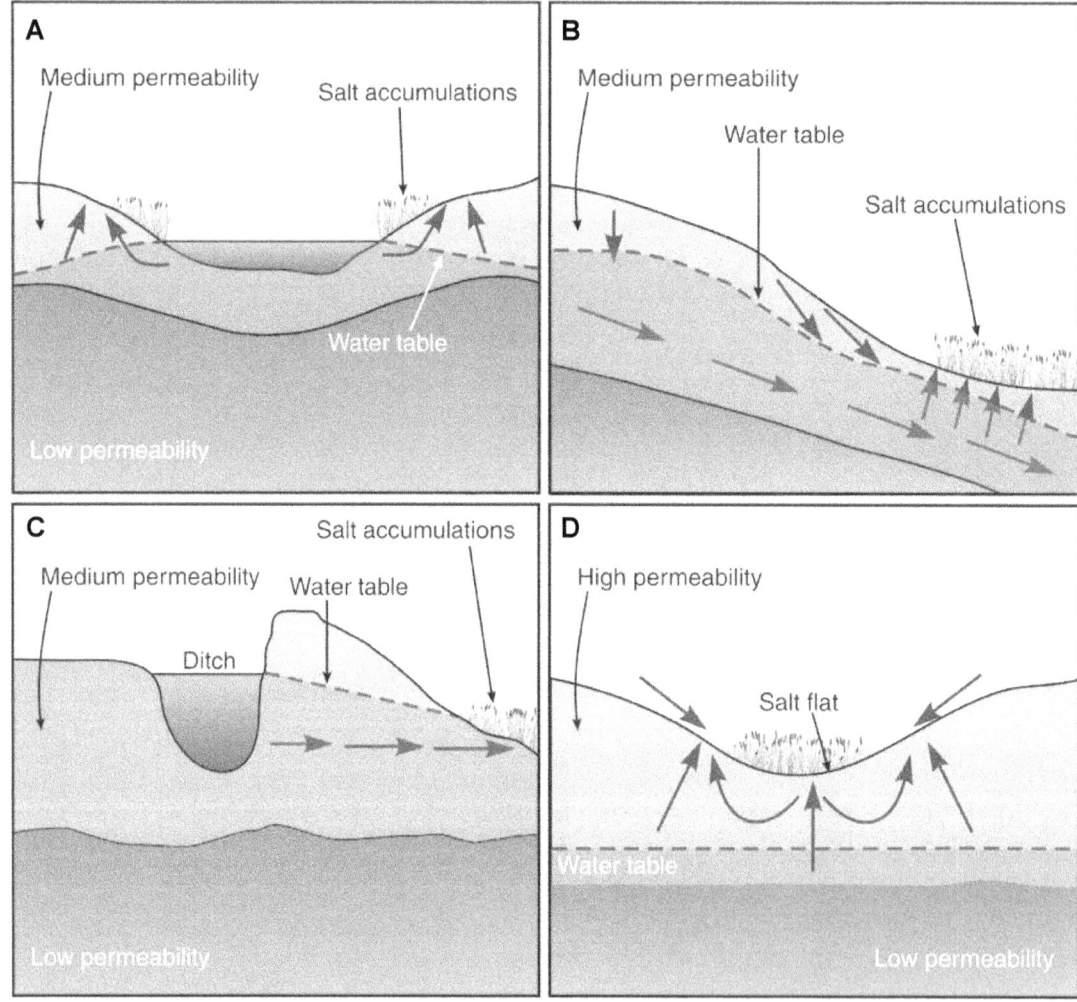

Figure 2-10. Schematic illustrations of different hydrogeomorphic settings that create salt flats. Ponds create salt accumulation where they dry seasonally or where the surface water-supported water table is close to the soil surface (A). Salt also accumulates where groundwater discharges to the surface and evaporates (B and C), or where capillary water reaches the soil surface (D). Modified from Alberta Agriculture and Food (2004).

24

USDA Forest Service Gen. Tech. Rep. RMRS-GTR-282. 2012

One approach to the classification of wetlands, the hydrogeomorphic system, provides a classification based upon site properties (Brinson 1993). Indicators of function are discussed as derivatives of the three basic properties along with the ecological significance of each of the properties. The core of the classification has three components: (1) geomorphic setting, (2) water source and its transport, and (3) hydrodynamics. Geomorphic setting is the topographic location of the wetland within the surrounding landscape. Water sources can be simplified to three: precipitation, surface or near-surface flow, and groundwater discharge. Hydrodynamics refers to the direction of flow and strength of water movement within the wetland. While the three components are treated separately, there is considerable interdependency. Such redundancy may be useful where it reduces errors in interpretation and reinforces the underlying principles that explain wetland functions. As shown in Table 2-1 (Smith and others 1995), there are a number of different hydrogeomorphic classes, several of which occur in the western United States. Riverine wetlands occur along streams; depressional wetlands form in basins and typically are marshes or fens; and slope wetlands are groundwater discharge driven and support fens and wet meadows (Figures 2-10 and 2-11). Mineral soil flats include salt flats.

Table 2-1. Hydrogeomorphic classes of wetlands showing dominant water sources, hydrodynamics, and examples of subclasses (Brinson 1993).

Hydrogeomorphic class (geomorphic setting)	Water source (dominant)	Hydrodynamics (dominant)	Examples of regional subclass	
			Eastern USA	Western USA and Alaska
Riverine	Overbank flow from channel	Unidirectional, horizontal	Bottomland hardwood forests	Riparian forested wetlands
Depressional	Return flow from groundwater and interflow	Vertical	Prairie pothole marshes	California vernal pools
Slope	Return flow from groundwater	Unidirectional, horizontal	Fens	Avalanche chutes
Mineral soil flats	Precipitation	Vertical	Wet pine flatwoods	Large playas
Organic soil flats	Precipitation	Vertical	Peat bogs; portions of Everglades	Peat bogs
Estuarine fringe	Overbank flow from estuary	Bidirectional, horizontal	Chesapeake Bay marshes	San Francisco Bay
Lacustrine fringe	Overbank flow from lake	Bidirectional, horizontal	Great Lakes marshes	Flathead Lake marshes

USDA Forest Service Gen. Tech. Rep. RMRS-GTR-282. 2012

25

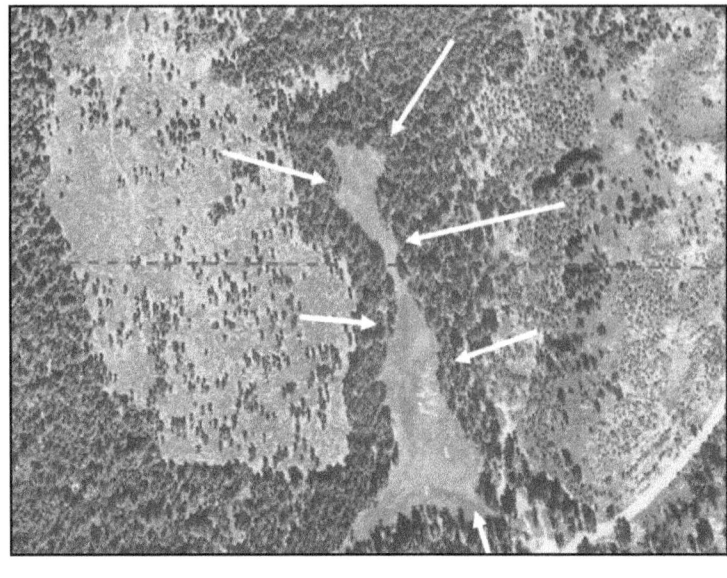

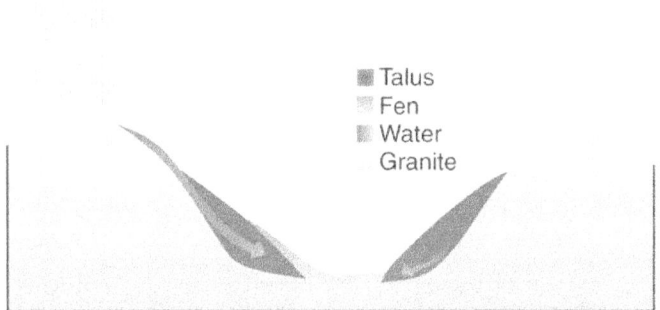

■ Talus
▨ Fen
■ Water
　Granite

Figure 2-11. (Top) Aerial photograph of Poison fen, Sierra National Forest, California, showing the directions of groundwater flow from granite domes, unforested uplands on east (bottom) and west (left) side of the fen. Groundwater discharges into the fen and sheet flows to the south, as shown with red arrows. (Bottom) Illustrative cross section along the dotted blue line (in top frame) showing the granite domes, hillslope talus and alluvium, and groundwater flow and discharge patterns that form the sloping Poison fen (green areas) at the toe of slope.

Landscape form and function

Hydrologic and geomorphic processes determine where wetland and riparian ecosystems occur in landscapes, and they control the type of wetland that occurs as well as the way it functions. The wetland ecosystem types that occur in the western United States (fens, marshes, salt flats, riparian areas, and wet meadows) can be defined by their hydrologic and geomorphic regime. Hydrologic processes supporting wetlands and riparian areas can be complex. Many wetland complexes have multiple surface and/or groundwater flow systems (Winter and others 2001). Wetlands dominated by groundwater versus surface water and precipitation may respond differently to natural and anthropogenically driven climatic events and variability (Winter 1999). Wetlands that depend primarily on runoff from precipitation for their water supply, such as some marshes, may have highly variable water levels, while those dependent on discharge from local or regional groundwater flow systems, such as most fens, have the least variable water levels (Winter 1999, 2001).

Geomorphic processes drive many disturbances, and landform generation strongly influences hydrologic processes and vegetation patterns in all wetland types. For example, riparian landforms are generated by flood-driven sediment erosion and deposition that leads to channel changes and creates a mosaic of landforms such as channels, floodplains, point bars, and in-channel islands (Gregory and others 1991) (Figure 2-12). These landforms influence the spatial pattern and successional development of riparian vegetation. Riparian plant establishment is linked to the frequency and magnitude of flood driven landscape disturbances that produce bare and moist sediment (Baker 1990, Auble and others 1994, Scott and others 1997, Johnson 2000, Friedman and Lee 2002, Cooper and others 2003a). For example, braided streams have very high erosion and deposition rates and are highly dynamic. Ephemeral streams have only periodic flows; little vegetation may occur in the channel or on the floodplain; and the stream may wander across the floodplain on a time scale of hours or days. Stream avulsion may occur when meander bends erode to shorten a channel length, thereby forming a new channel (Figure 2-12), or when a beaver dam pushes flow across the floodplain. Disturbance regimes vary as a function of position in the landscape, and the distinctive style, magnitude, frequency, and duration of disturbances may be categorized into process domains (Montgomery 1999).

At a landscape scale, glaciers have created the template for wetland formation in many regions where glacial till has blocked drainages and created large bodies of unconsolidated sediments as well as landforms such as kettle basins and moraine-dammed basins. This sediment is recharged with groundwater and may form aquifers with important groundwater flow systems. Glaciers originating in high mountain cirques have created broad, U-shaped valleys instead of the V-shaped valleys characteristic of unglaciated areas.

Identifying the water sources and hydrologic regimes of wetlands and riparian areas is the first step in understanding the ecological patterns and processes occurring at a site of interest. Putting a site into context by identifying its hydrogeomorphic class and its process domain can also assist in identifying appropriate reference sites. Such context may assist in assessing desired conditions and in identifying sources of stress that may be modifying processes that maintain the functions that a wetland or riparian area performs. Furthermore, understanding natural disturbance regimes and distinguishing them from human-caused disturbances can help in determining appropriate management actions that facilitate desired functions of a site.

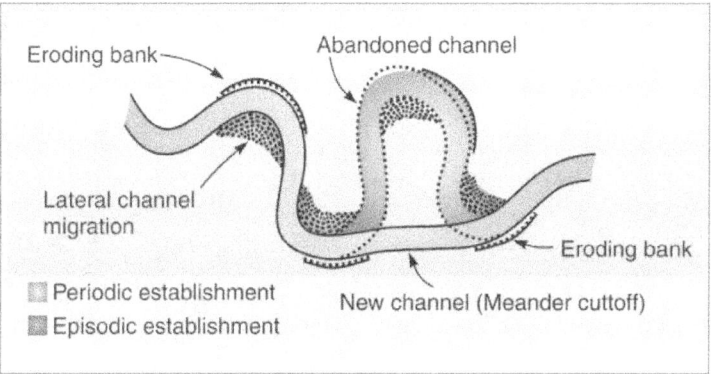

Figure 2-12. Processes of lateral channel migration and meander cutoff (modified from Richter and Richter 2000). Establishment of riparian cottonwoods and willows occur as a consequence of channel migration and point bar formation (green areas) as establishment sites are created by stream bank erosion on the outside of meander bends and the deposition of transported sediments on point bars. Channel abandonment and beaver pond formation and failure may create large areas suitable for riparian plant establishment, allowing for the formation of large cohorts of seedlings. From Cooper and others (2006) and Westbrook and others (2006).

Chapter 3: Plant Water Requirements and Vegetation Sampling

How plants acquire and use water

Terrestrial plants acquire water to obtain mineral ions and nutrients from the soil, to maintain cell turgor pressure and prevent wilting, and to fix carbon through the process of photosynthesis. Vascular plants have specialized tissues for water absorption and transport, including roots for absorption, xylem vessels (or tracheids) for transport, and leaves that regulate water loss. Vascular plant cells have a large central vacuole and maintain hydration via water acquisition primarily from the soil. Non-vascular plants, such as bryophytes and lichens, lack specialized tissue for water transport, so they must directly acquire atmospheric moisture (humidity, dew, and precipitation) or near surface soil water. Bryophytes and lichens may periodically be dry or only rarely wet, and their water content typically changes seasonally.

A prerequisite for plant photosynthesis is gaseous exchange with the atmosphere through specialized openings in leaves called stomata. Stomata allow carbon dioxide (CO_2) to enter the leaves, providing a carbon source for making carbohydrates. However, when stomata open to allow CO_2 into the leaf, water may escape into the atmosphere in a process called transpiration. The rate of water loss is driven by atmospheric demand for water, measured as a vapor pressure deficit (VPD)—the difference between how much water a parcel of air can hold at a given temperature and the actual amount of water vapor present. VPD drives the energy gradient that pulls water up through the plant, creating the soil-plant-atmosphere continuum. The difference between atmospheric demand for water and moisture availability to the roots influences the internal water status, or water potential, of the plant. Transpiration may create stress on plants if water is not available in the soil to replace that which is lost to the atmosphere. The plant may respond by regulating (constricting or closing) its stomata, if possible. Prolonged water deficit may result in leaf wilting, leaf death, xylem cavitation, branch loss, and, ultimately, whole plant death (Rood and others 2000).

Stomatal resistance is the most significant regulation of transpiration for plants that closely regulate their stomata because it occurs at the steepest part of the water potential gradient, where open stomata expose the interior leaf cells to the potentially desiccating atmosphere. Guard cells adjust the stomatal opening to reduce water loss when soil water availability is limited or when very high atmospheric demand makes it impossible for a plant's vascular tissue to provide sufficient water to leaves. Most species undergo daily or seasonal stomatal adjustments in response to water availability. Stomata in most plants close at night to reduce transpiration when the lack of sunlight reduces the need for CO_2 exchange for photosynthesis. This stomatal control allows many species to occur in environments that experience seasonal or nearly permanent drought conditions or extreme fluctuations in water availability and/or atmospheric conditions. Stomatal aperture, transpiration rates per unit leaf area, and photosynthetic rates per unit leaf area can all be measured and provide an excellent means of understanding plant functioning.

Vascular plants often have well-developed root systems that allow them to acquire water from a range of soil depths and, in some cases, directly from the saturated zone (surface or groundwater) or the capillary fringe. Plants that acquire water from the water table and its capillary fringe are termed phreatophytes. Vascular plants typically have

USDA Forest Service Gen. Tech. Rep. RMRS-GTR-282. 2012

29

greater stomatal control than non-vascular plants and dominate a wide range of environments, including deserts, tropical forests, and polar highlands. The only environments dominated by non-vascular plants are constantly moist or wet environments such as bogs and some fens (peatlands). Sites that are seasonally or periodically wetted, including wet meadows, many riparian areas, and all uplands, may have little moss and lichen cover because non-vascular plants lack roots and have little ability to regulate their stomata.

Water enters vascular plants primarily through root hairs (Figure 3-1) that have great surface area for absorption and high permeability to water. Once water enters the root, it flows to vascular tissue where it moves up stems to leaves. Water flows along a pressure gradient, which is greatest between the atmosphere and leaf and lowest from soil to root. The trunks of woody plants such as trees and shrubs have several tissues, including bark and phloem that moves food from photosynthetic to non-photosynthetic portions of the plants, and the cambium, where phloem and xylem cells are formed (Figure 3-2). The xylem or wood is composed of sapwood that actively transports water and heartwood, which provides structure but little water conduction. Most xylem tissue is composed of dead hollow cells with lignified walls.

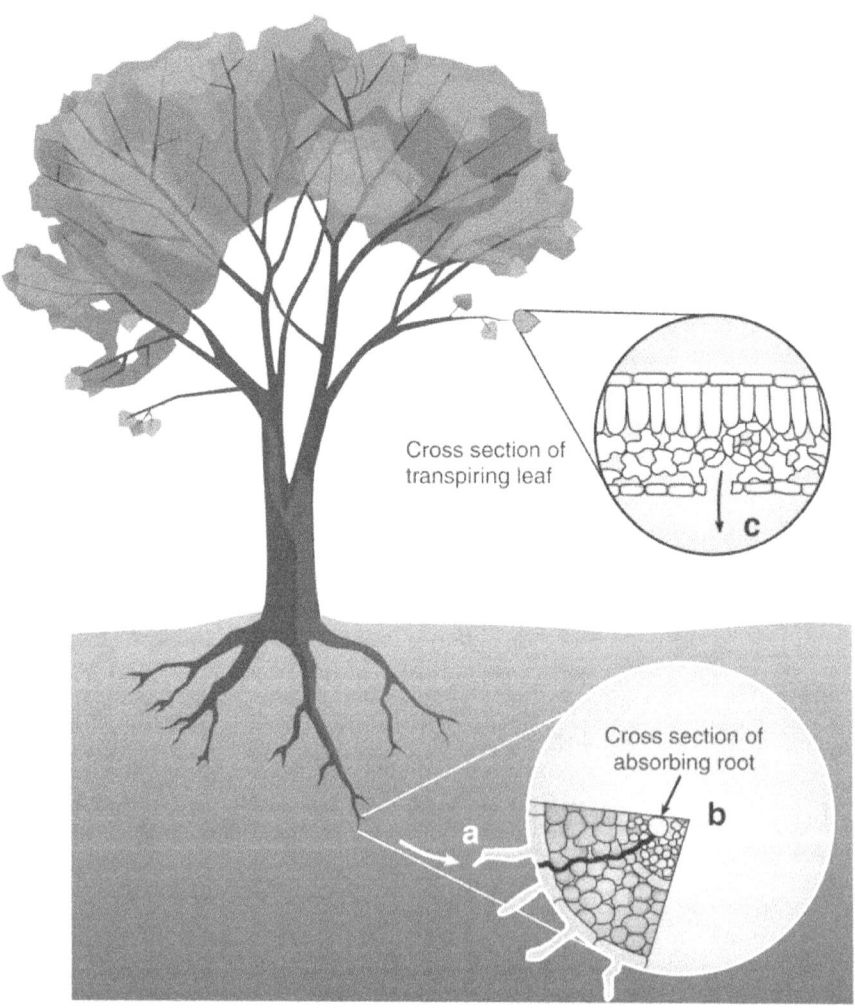

Cross section of transpiring leaf

c

Cross section of absorbing root

a

b

Figure 3-1. Whole tree showing a cross section of leaf with stomata (c), and a cross section of a root showing rootlets and root cells (b).

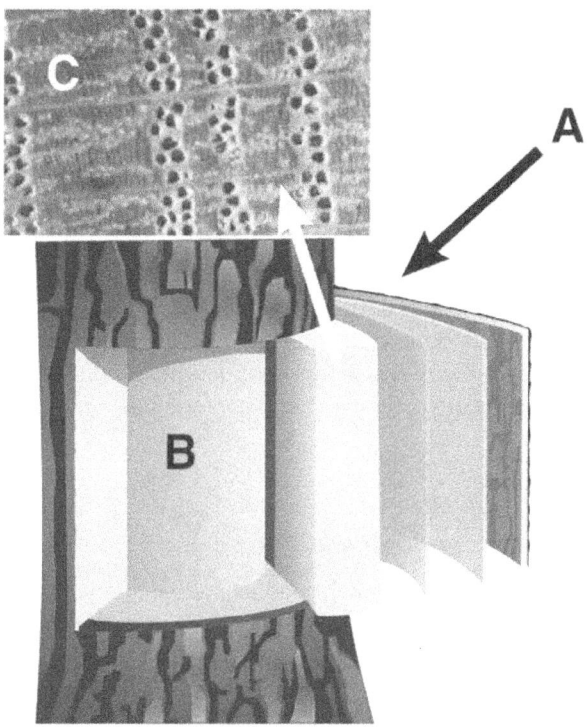

Figure 3-2. Cross section of typical tree showing layers of woody tissue. The arrow at (A) points to the outer layer, which is bark. Inside the bark in succession are the phloem, cambium, sapwood, and heartwood shown at (B). The sapwood is the main water conducting tissue in the tree and contains xylem vessels. Large pores are shown in (C).

Many species of riparian plants require access to a permanent or seasonal water supply and are intolerant of low internal water potentials (e.g., high water stress). Whereas most phreatophytes are thought to maintain contact with the vadose zone (unsaturated zone of capillary rise above the water table), even during periods of low flow, other species may have an affinity for fine textured substrates with high water holding capacity or may be able to utilize water at low soil water pressure potentials (Naumburg and others 2005). Some riparian species can utilize different water sources over the course of the season, with various proportions of transpired water coming from groundwater versus soil water depending on relative availability (Busch and Smith 1995, Smith and others 1998). Species that can acquire water from multiple sources may be better adapted to extended periods of low flow caused by drought, groundwater pumping, and water extraction (Stromberg and others 2007a).

Flow permanence (length of time with surface flow) and depth to groundwater, accounted for much of the variability in dominance by native or exotic species along a riparian hydrologic gradient in Arizona, USA (Lite and Stromberg 2005). Sites that had surface flow for 76 percent of the time, groundwater depths of <2.6 m, and variability of groundwater depth of <0.5 m tended to be dominated by native forest species. Decreasing flow permanence and increasing depth and variability of groundwater beyond these thresholds resulted in an increasing probability that native-dominated woodlands could shift to exotic shrublands, leading to changes in stand-level vegetation characteristics (e.g., canopy height and vegetation volume). Along a depth to water table gradient, non-riparian phreatophytes, such as *Sarcobatus vermiculatus* (greasewood), also vary their use of soil and groundwater (Chimner and Cooper 2004).

When subjected to prolonged periods of drought, even the most drought-tolerant riparian plants are vulnerable to water stress, wilting, leaf death and stem dieback, and, ultimately, mortality (Tyree and others 1994, Scott and others 1999). As desiccation occurs, species vary greatly in their ability to regulate their water pressure potential (through stomatal closure or other leaf morphologic traits) (Pockman and Sperry 2000). Loss of leaves through leaf abscission and/or xylem cavitation and branch death ("drought pruning"), reduces the total plant water requirements and may actually save the individual at the cost of some portion of the canopy (Tyree and others 1994). For some species, such as *Populus deltoides*, there is an identifiable threshold of percent leaf and branch loss before mortality is likely (Scott and others 1999, Cooper and others 2003b).

Measuring plant water relations

Stomatal conductance

Measures of stomatal conductance are important for understanding transpiration rates and water use efficiency. Furthermore, the diurnal and seasonal patterns of plant adjustment to changes in water availability provide key information for understanding plant functioning. For example, the diurnal pattern of stomatal conductance (g_s) is important for understanding how plants function during the day and recover at night (Oren and Sperry 1999).

Leaf stomatal conductance and transpiration rates can be measured with a porometer or portable photosynthesis system (Figures 3-3A and B). This instrument measures the rate of dry air needed to maintain a constant relative humidity inside a small chamber enclosing a transpiring leaf. Photosynthesis systems utilize one or more infra-red gas analyzers that measure the CO_2 and H_2O vapor concentrations within the leaf. Porometer measurements are made on leaves *in situ* (Figure 3-3A). Ladders or scaffolding may be needed to access suitable leaves. Leaves or twigs that hold the set of leaves for measurement should be marked, making repeat measures over the day or season possible. Data

Figure 3-3. The measurement of transpiration and stomatal conductance of a *Populus* spp. tree using a porometer (Li-COR LI-1600 or similar instrument).

depicted in Figure 3-4 are from an experiment where *Populus* spp. trees were irrigated and compared with unwatered trees. Plants opened their stomata at dawn and reached maximum g_s by early to mid morning (Figure 3-4). By late morning, unwatered plants were adjusting their stomata to reduce transpiration, while watered plants were not. On a seasonal basis, stomatal conductance may increase, however morning g_s is higher than afternoon g_s throughout the summer, indicating that afternoon stomatal control occurred due to limited water availability and a desiccating summer environment.

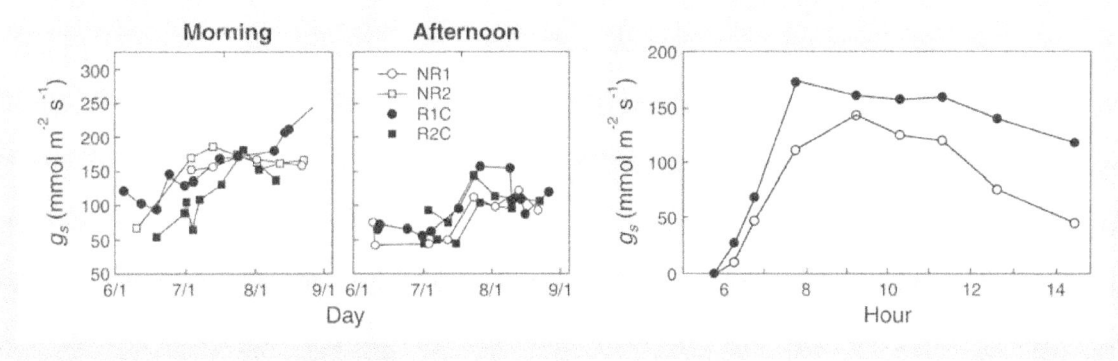

Figure 3-4. Morning and afternoon stomatal conductance (left panels) and diurnal patterns of stomatal conductance of *Populus* spp. trees using a porometer. Irrigated *Populus* are indicated with black filled symbols; unwatered *Populus* are unfilled symbols at sites NR1, NR2, R1C, and R2C.

Xylem pressure potential

Water is a polar molecule. As a result, water molecules adhere to each other and to the vessel walls of the xylem. Transpiration pulls the chain of water molecules up the plant, but when water is transpired faster than it can be replaced in the xylem, tension on the chain of water molecules in the xylem increases and is exerted through the entire plant. When a twig or leaf petiole and its xylem is severed from the parent plant, the pressure necessary to force water back to the cut surface is equivalent to the negative pressure (internal water stress) in the xylem prior to cutting. A Scholander-type pressure chamber can be used to measure this pressure and provides an excellent measure of tension within the xylem. A leaf with a petiole or twig is cut and placed upside down into the chamber with the cut end protruding (Figures 3-5). Pressure is increased in the chamber and the cut end is observed with a magnifying glass. When the pressure exerted equals the tension at which water is held within the xylem, water flows back through the xylem and appears on the cut end. When water first appears, this xylem pressure potential is recorded and represented with the symbol Ψ_{xp}.

Differences in hydraulic architecture make some tree species more sensitive to water stress than others (Sperry 1995, 2000). Vulnerability curves have been developed for a number of species to quantify their loss of hydraulic conductivity along a gradient of xylem pressures (Tyree and Sperry 1989). For example, Tyree and others (1994) produced vulnerability curves for *Populus* spp. to show loss of water conducting ability and the pressures at which partial or complete loss of xylem conductance occur (Blake and others 1996).

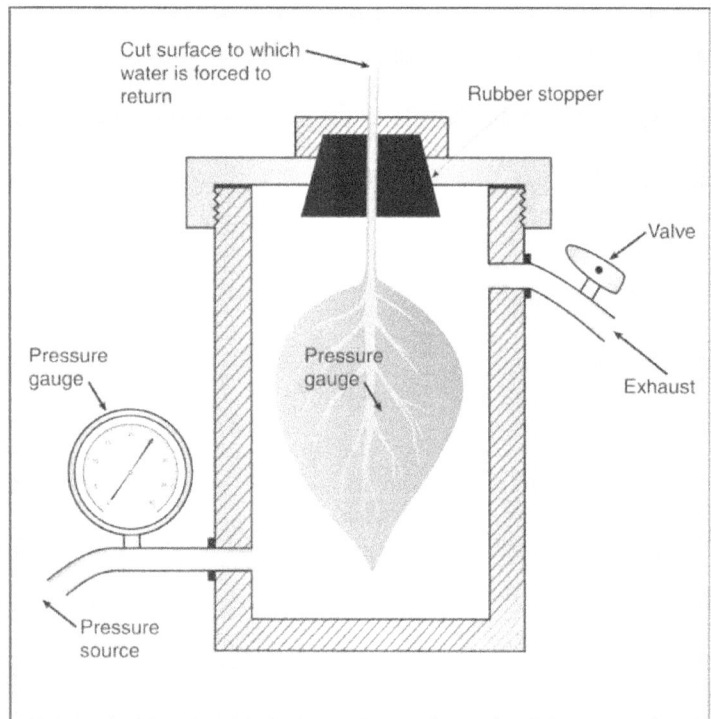

Figure 3-5. Left: cross section of pressure bomb showing cut leaf (or twig) inside the chamber with the cut end protruding through the rubber stopper and chamber top. The exhaust valve is closed and pressure increased until water flows back through the xylem and is expressed on the cut leaf or twig surface (top right). The researcher carefully watches the cut surface as the pressure increases, recording the pressure at which the water emerges (bottom right).

Measuring transpiration rates

Transpiration is an important determinant of leaf energy balance and plant water status (Pearcy and others 1989). Most studies of plant-water relations involve measurement of leaf transpiration and leaf conductance to water vapor as they enable an investigator to determine plant water use efficiency. Transpiration can be measured at the scale of the leaf, individual plant, or a meadow, forest, or stand using a variety of techniques. These measurements require specialized equipment and expert knowledge of the techniques and theory behind them, so clear objectives should be made in the study planning phase. Brief explanations of several techniques follow. For more detailed explanations, theory, and instructions, please refer to plant physiology textbooks (e.g., Chapters 3, 8, 9, 11, and 13 of Pearcy and others 1998) and/or instruction manuals for specific measurement devices.

Leaf-level transpiration is typically measured as previously described. Leaf-level water conductance (water exiting a leaf through stomata) is often derived from transpiration rates measured in chambers. A number of companies make instruments for measuring leaf-level water conductance in the field, including LI-COR (www.licor.com), PP systems (www.ppsystems.com), and Decagon (www.decagon.com). Leaf-level measurements may be made on multiple leaves on the individual of interest over the course of a day (or season) to better understand the physiological response of the plant to changes in air temperature, humidity, precipitation events, soil moisture, water levels, and streamflow.

Whole plant-level transpiration rates may be analyzed using heat as a tracer to measure water flux through the stems of woody plants. One method for measurement of sap (water) flux through trees uses pairs of probes inserted through the sapwood as shown in Figures 3-6A, B, and C. The bottom probe measures ambient temperature of the wood. The top probe measures temperature and also has a heating element that produces a constant flow of heat. Water conducts heat and transports it up and away from the heating element. The difference in temperature between the two probes is measured. Higher temperature differences between the two probes indicates a slow flow of water up the tree, while a smaller difference indicates a higher rate of flow. Water movement up the tree is calculated as sapwood cross sectional area multiplied by the rate of sap flux. These data can be calculated as daily or annual flux. The thickness of sapwood for a tree (light area in the cross section shown in Figure 3-6A) can be determined by measuring sapwood thickness with increment cores (Figure 3-6B), or other methods, from around the tree bole or through measurement from a cross section of the stem.

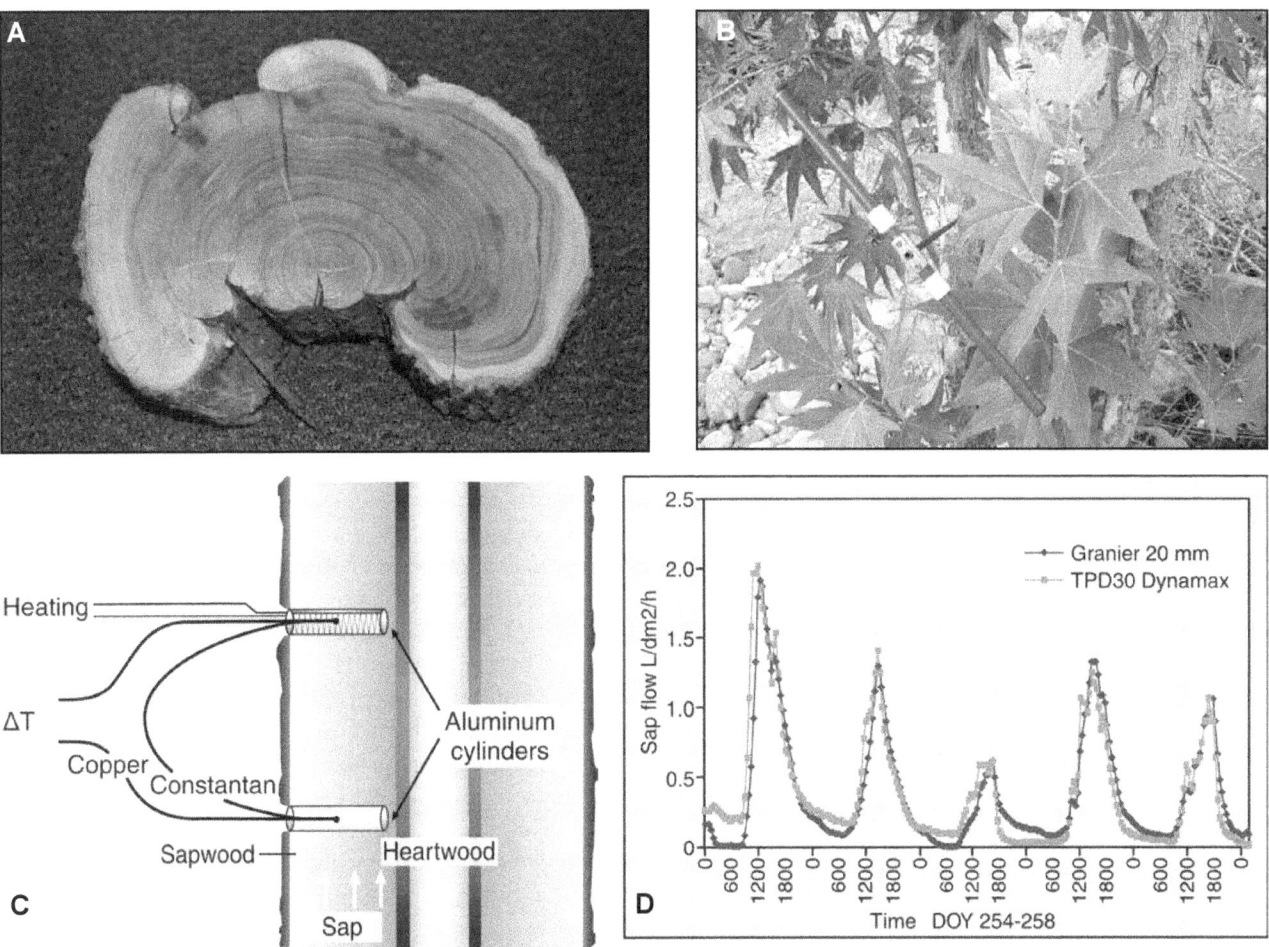

Figure 3-6. Sap flow methods can be used to calculate flow rates of water up the sapwood of tree boles. Knowledge of the sapwood thickness and area (A and B) is determined with cores, or cross sections, and flow rates are determined with pairs of probes (C) or bands. Sap flow can provide informative (D) diurnal patterns of water use in trees. The Granier 20 and TPD30 are two different types of sap flow probes, and a number of methods and systems are available for sap flow studies.

However, there are many new approaches for measuring sap flux and many different probe types are available, each of which may be most appropriate for a particular tree species or question. Since differentiation of active xylem can be challenging for some species, laboratory analyses are required and probes can be calibrated in the lab following the procedures of Steppe and others (2010). Sap flux technology is complex and a thorough understanding of the methods, calibration, and analysis of these data is required to successfully utilize these techniques. Many studies have shown that the original Granier sap flow equations are not robust for all species, and new ideas about how to measure sap flow are being published (Clearwater and others 1999, Taneda and Sperry 2008, Bush and others 2010, Hubbard and others 2010). Sap flux over time may be plotted and compared across treatments, through space, and over time (Figure 3-6D).

Identifying plant water sources

Many wetland and riparian plants utilize groundwater as well as soil water that is recharged by precipitation. Although many wetland and riparian plant species are phreatophytes, they may primarily utilize soil water recharged by precipitation. Some plants are adept at taking advantage of groundwater and soil water at different times of the year (Busch and Smith 1995). Understanding the water sources used by plants is critical to understanding plants' link to, and degree of dependency upon, groundwater. Relationships between incremental growth, branch growth, productivity, and canopy condition and hydrologic variables (such as streamflow) can provide strong clues about linkages between fitness and various water sources to plants (Stromberg and Patten 1991, Willms and others 1998). In determining water sources and needs for riparian vegetation, it is important to understand the relationship between plant age or developmental stage, root morphology, and water acquisition. Vulnerability to water stress may decline as a function of age or developmental stage for many species. As a result, it is important to understand how many years it takes for woody plant seedlings or saplings to develop roots deep enough to acquire groundwater in the summer, or to determine the proportion of rain-recharged soil water that typical phreatophytes utilize.

Stable isotope ratios of oxygen (O) and hydrogen (H) can be used as tracers to identify water sources. Water is composed of O and H, and both elements occur in elemental and isotopic forms. O has an atomic number (protons) of 8 and an atomic mass (neutrons + protons) of 16. However, its most common isotope has two extra neutrons, and while its atomic number remains identical, its atomic mass is 18. H has an atomic number and atomic mass of 1; however, its most common isotope deuterium (D) has an atomic mass of 2. If a basin of water is set in the sun, water molecules with D or ^{18}O evaporate more slowly (due to their greater atomic mass) than do water molecules with H or ^{16}O, leading to an enrichment of water in D and ^{18}O. Evaporative enrichment occurs in most soils that are recharged by precipitation. The ratio of D to H (expressed as δD), or $^{18}O/^{16}O$ (expressed as $\delta^{18}O$) can be measured with a mass spectrometer. If δD or $\delta^{18}O$ is distinct in soil water as compared to groundwater, then the water source used by plants can be identified by comparing plant sap extracted from suberized stems (those with well-developed bark) with potential water sources. Water is extracted from soil samples and plant stems using a process called cryogenic distillation, and groundwater is collected by pumping water from monitoring wells. The isotope ratio is then calculated relative to a standard water source using the following equation:

$$\delta D\ (\%_o) = [(D/H)sample/(D/H)standard\ \text{-}1] \times 1000$$

The typical standard is Standard Mean Ocean Water (SMOW).

Cooper and others (1999) used stable isotope methods to determine that *Populus* seedlings along the Yampa River were using primarily soil water during the summer until they were at least two to three years old (Figure 3-7). Older plants used primarily groundwater, although much older plants (>90 years old) used both soil and groundwater. This investigation identified that seedlings were not phreatophytes and that their survival was not dependent upon growing a taproot fast enough to remain connected to the declining summer water table. Using stable isotopes, Busch and others (1992) found that a group of native riparian woody species (including *Populus* spp. and *Salix* spp.) were obligate phreatophytes along an arid land river but that a non-native shrub utilized both soil and groundwater during the growing season.

Stable isotopes of carbon (C) can also be used for analyzing plant-water relations (Jackson and others 1993). The isotopic ratio of ^{13}C to ^{12}C in plant tissue is less than the isotopic ratio of ^{13}C to ^{12}C in the atmosphere, indicating that plants discriminate against ^{13}C during photosynthesis. Variation in discrimination against ^{13}C is due to both stomatal limitations and enzymatic processes. Theoretical and empirical studies have demonstrated that carbon isotope discrimination is highly correlated with plant water use efficiency, providing an integrated measure of water use efficiency (a measure of carbon fixation for an amount of water transpired). Measurement of carbon isotope discrimination is relatively easy to carry out because carbon used to build plant tissue reflects the amount of discrimination present when the tissue was constructed. Samples are easily collected in the field for later processing in a laboratory. Moreover, in woody plants, carbon isotope discrimination can be determined from annual ring samples, providing a historical analysis of plant response to environmental conditions. Using carbon isotope discrimination to determine water use efficiency, Busch and Smith (1995) found that a non-native riparian species had higher water use efficiency than several native shrubs. This higher water use efficiency

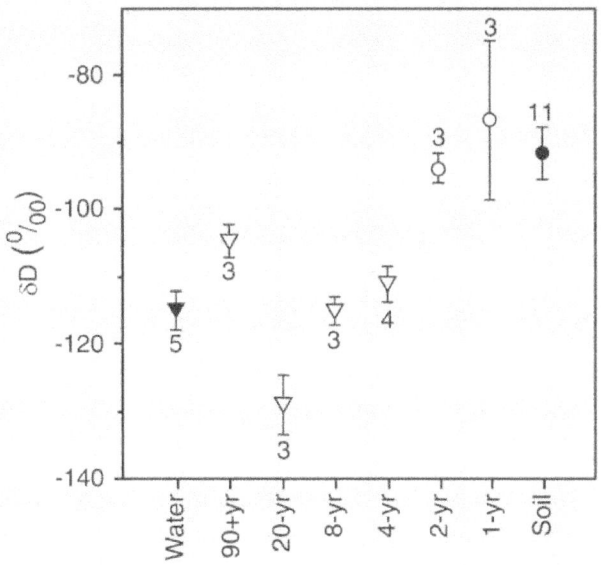

Figure 3-7. Sap water δD of groundwater (water); plants of 90+, 20-, 8-, 4-, 2-, and 1-year old cottonwoods; and soil water, illustrating that younger plants used primarily soil water.

provided an explanation for the differential survival of non-native versus native species along a flow altered river. Studies have found that depleted groundwater and altered timing and magnitude of flow may cause a shift in riparian areas from those dominated by native obligate phreatophytes to those dominated by non-native species (Stromberg and others 2007b, Merritt and Poff 2010).

Hydrogen isotope composition of cellulose from tree rings may also be used as an index of historical water source use by riparian trees (Alstad and others 2008). Analysis of cellulose D provides a time-integrated signal of changes in water sources to woody riparian plant species (currently only about a decade into the past) and may provide indications of switches in water use from one source to another or may provide insight into reductions in water availability from an existing source. For example, analysis of cellulose D from tree ring tissue may provide insight into a switch in water use from groundwater to soil or atmospheric water following groundwater decline or streamflow alteration.

Sampling of tissue and water for stable isotope analysis involves collection, proper handling, and timely delivery of samples to a laboratory for analysis using an isotope ratio mass spectrometer. Collection of water samples from groundwater wells, rain gauges, stream water, and/or soil water as well as water in plant tissues may be necessary for determination of isotopic composition, depending on the objectives of the study. Water samples should be immediately sealed in filled glass vials to avoid any evaporation of water from plant tissues. Leaves, stems, and xylem water from tissue samples have commonly been used in stable isotope studies and may be collected by clipping or pruning the desired portion of the plant. Samples should be packaged and chilled (preferably frozen) and shipped immediately to the laboratory to avoid fractionation and loss of organic material. For a more thorough discussion, refer to Ehleringer and Osmond (1989).

Laboratories that perform isotopic analysis on tissue and water samples include:

University of Colorado: http://instaar.colorado.edu/sil/about/index.php;
University of California, Davis: http://stableisotopefacility.ucdavis.edu/;
Northern Arizona University: http://www.mpcer.nau.edu/isotopelab/pricing.html;
and many private laboratories.

Plant ages and growth

A critical issue in riparian management in western North America is *Populus* and/or *Salix* establishment and survival. Since rivers have varying interannual flow regimes, knowing the exact year woody plants establish allows researchers to link plant establishment to a particular year's flow regime. In addition, determining the age structure of woody plant populations is critical for helping to understand the years in which plant establishment success has been high, the hydrologic conditions associated with establishment and survival, and possible hydrologic or climatic bottlenecks in survival of such species. The determination can also allow scientists to suggest flows for environmental maintenance or restoration.

When the seed of a woody plant such as *Populus* germinates, its cotyledon grows a stem and a root. This germination point, which occurs at the soil surface (Figures 3-8 and 3-9), is termed the root crown or root collar. As the plant grows and develops woody tissue, stem tissue develops a pith. Stem tissue can always be identified in cross section by the presence of pith (even following burial), while root tissue lacks a pith.

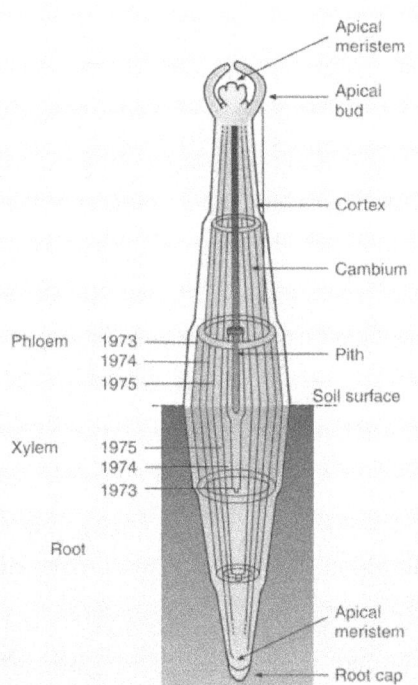

Figure 3-8. Diagram of young tree showing the soil surface and the stem tissue above and root below that point. The stem contains a pith and growth rings reflecting each year's growth. The point where the pith originates is the root crown.

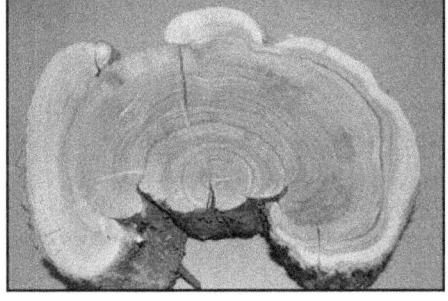

Figure 3-9. Photograph of an excavated *Tamarix*. The sanded *Tamarix* spp. slab at bottom contains a pith (the light center), indicating that it is stem tissue. Root tissue lacks a pith. Excavated *Populus* bottom right.

USDA Forest Service Gen. Tech. Rep. RMRS-GTR-282. 2012

39

Because stems grow upward and add successive layers of wood in a ring-like pattern, the younger stem areas contain fewer rings than the root crown. While an increment core collected above the ground surface from a tree bole can be used to obtain a general idea of tree age, it cannot be used to determine the year in which riparian tree establishment occurred (Figure 3-10). Along fluvially active streams, the stems of individuals may be sheared off or buried by sediment, causing the germination surface of the resprouted plant to be buried and the aboveground portion of the plant not to reflect the true age of the individual. When attempting to determine establishment age of such individuals, the plant must be excavated to find the root crown, and the age at the root crown must be determined through examining growth rings. For shrubs, or browsed plants, it should be assumed that no stem contains the full set of rings, and excavation of the plant to identify the root crown is necessary. Through excavation of the roots of willow, cottonwood, or *Tamarix* (Figure 3-9), root morphology can be characterized and the taproot can be preliminarily identified. Then, cross sections of the stem should be collected, and the root crown can be identified. The root crown is located between the cross section that has pith (stem tissue) and the cross section that does not contain pith (root tissue). If a cross section has pith on the top and no pith on the bottom, the root crown (and germination surface) is contained in the cross section.

This section bridges the root crown and should contain the full complement of annual rings. If plants are deeply buried by fluvial sediments, stem tissue becomes root tissue, which can be very porous and, in some cases, difficult or impossible to interpret. In addition, trees along some intermittent streams or following drought may produce false rings (more than one per year). In extreme cases, rings may be absent. Therefore, it is important to cross date ring widths among trees and relate them to known climate and hydrologic events.

In addition to simple ring counts to age trees, cores extracted from aboveground stems may also be analyzed to measure the width of incremental (annual) growth rings and to quantify tree growth rates. Tree cores may be measured precisely using a binocular microscope. Ring width may be digitized using specialized equipment such as a Velmex TA Unislide measuring system with an ACU-Rite linear encoder and QC1100 digital readout device (Velmex, Inc., Bloomfield, New York). Ring-reading software

Figure 3-10. An increment borer may be used to collect cores of trees and shrubs. Cores are then dried, mounted onto a board, sanded, and polished. Rings are counted or ring widths are measured to determine age or growth rate, respectively.

includes Measure J2X, Version 3.1 (Project J2X, Voortech Consulting, Holderness, New Hampshire).

Another growth measure that is tightly coupled with water availability in arid and semiarid riparian systems is annual branch growth. Measuring the increments between annual bud scale scars provides a measure of growth in the past growing seasons. An annual branch growth measurement may be sensitive (in some cases, more so than incremental ring growth) to the effects of changes in groundwater levels (e.g., groundwater pumping; Scott and others 1999) and to variations in river flow regime (Willms and others 1998). Branch growth increments may provide an accurate record of environmental favorability for recent growth over a period of one to two decades. Close correlations between branch growth and stream flow indicate that water limits growth of riparian plants and that such plants obtain their water from a source linked to the stream, such as the riparian water table (Willms and others 1998). Analyses of branch increments provide a management tool for determining instream flow needs for riparian trees and for analyzing impacts of stream flow alterations due to river damming, water diversion, and groundwater pumping.

Patterns of plant dieback

The outward signs of water stress in plants may include discoloration, wilting, and/or dead leaves or branches. Woody plant stems or root systems may dieback due to drought, water management actions, disease, and other factors. Understanding the patterns of dieback assists in linking plant health to environmental and management actions.

Other plant morphological measures that can be useful in assessing riparian and wetland health and tracking changes in condition through time are: vegetation volume, canopy cover, canopy height, woody plant stem density, and woody plant basal area (Stromberg and Patten 1991, 1996; Lite and Stromberg 2005). Vegetation volume may be measured using the vertical line intercept method at a number of points per plot (Mills and others 1991). Maximum canopy height within plots may be measured using either a vertical measuring pole or a clinometer. Canopy cover may be measured using a spherical densiometer at several points (plot corners and center) per plot or by using a densiometer with point or line intercept layout (Elzinga and others 2001).

Percent of potential canopy can be used to assess damage caused by water stress associated with leaf death and abscission, water stress and cavitation, and branch dieback (Scott and others 1999) (Figure 3-11). Potential canopy should be estimated as a visual determination of percentage of live canopy relative to potential crown volume (i.e., extent of all branches; Scott and others 1999) for all woody species. Crown dieback has been associated with increased risk of mortality the year following dieback in riparian trees (Scott and others 1999).

Root density and biomass with depth can be quantified from samples collected in pits or cores to compare areas affected by water management activities to reference areas. Williams and Cooper (2005) found that cottonwood root density and biomass was much higher for the unregulated Yampa River, which still has overbank flooding compared with the regulated Green River which rarely floods.

Dead stems can also be collected and their ring patterns can be compared with other live trees to identify the year of death (Figure 3-11). This requires the development of a tree ring chronology from healthy stems using collected increment cores. Analysis of tree populations in a holistic manner (evaluating their physiological functioning), including analysis of stomatal conductance and/or xylem pressure, age structure of stands, and dieback patterns of roots or stems, informs researchers on tree- and stand-level condition and the effects of long-term management activities on the persistence of riparian vegetation (Figure 3-11).

Figure 3-11. Individual trees, such as the cottonwood shown here, can yield a wide range of data and information, including physiological measures (top right), increment cores to illustrate overall growth rates and patterns of stem dieback (center image), root distribution and density (top left), and root crown age (bottom left).

Vegetation sampling

Vegetation is sampled to characterize the current composition of plants occurring on a site and to develop a baseline for future vegetation analyses. A large number of methods have been proposed and used for vegetation analysis; for an overview of this topic, please consult a textbook (e.g., Mueller-Dombois and Ellenberg 1975, Elzinga and others 2001). Two of the most popular methods are plot-based and transect-based sampling. Each method has its strengths and weaknesses and can be used for distinct purposes. Plots can be centered around groundwater monitoring wells to characterize the composition of homogenous stands or patches of vegetation in different hydrologic settings, or they can be around monitoring wells with different long-term water table depths. Transects can be used to sample vegetation along environmental gradients or in large stands where a broad sample area is desired (Figure 3-12). It is critical that sample size of plots meets the criteria for minimal area, that is, the minimal sample area that adequately represents the community composition. The sum of sample area along transects should also meet this minimal area criterion.

Figure 3-12. A tape measure laid out to measure vegetation using the line intercept method, a transect-based method.

USDA Forest Service Gen. Tech. Rep. RMRS-GTR-282. 2012

43

Some sample procedures use a single plot, while others have nested plots or nested plots placed along transects. Different plot sizes are used to sample herbaceous and woody plants. For example, 1x2 m plots are used to sample the composition and cover of herbaceous vegetation and two- to four-year old saplings (woody species), and 2x10 m plots are used to sample shrubs and trees (for estimation of shrub cover, tree stem density, and tree diameter). Along one transect, an example of plot layout is: 1x2 m plots located on the downstream side of the transect line/meter tape (the landward upstream corner of the plot on the selected distance along the tape). Each 2x10 m plot would have its origin (landward upstream corner) on the same point as the herbaceous plots. Each nested herbaceous and shrub-tree plot would correspond to its associated distance along one transect, with the exception of the belt transects nearest the lowest extent of vegetation, in which case each shrub-tree plot will be associated with two herbaceous plots.

If a single plot is used, the plot should be within a homogenous stand of vegetation. A list should be made of all plant species that are present, including bryophytes and other taxa if possible, and the absolute canopy cover of each species should be estimated.

Modified Braun-Blanquet cover classes are suitable for visually estimated vegetative cover in plots: cover class 1 = <1 percent canopy cover, 2 = 1 to 5 percent, 3 = 5 to 25 percent, 4 = 25 to 50 percent, 5 = 50 to 75 percent, and 6 = 75 to 100 percent. Within each plot, cover of each vascular species and ground cover feature (water, bare ground, litter, bryophytes, rock, and large wood pieces) should be recorded. If nested plots are used, or transects with different sized plots, the smaller plots are used to record the cover and abundance of herbaceous plants, bryophytes, shrubs, and young trees, and larger plots are used for trees.

Woody plant recruitment

The presence of woody species saplings should be recorded (by species) along transects when the stem or canopy of a two- to four-year old individual intersects the transect. Age may be determined by taking a cross section of several individuals off of the transect and counting annual growth rings. The presence of two- to four-year old individuals is considered evidence of recruitment, and the frequency of recruitment provides an important and sensitive measure of the recruitment success along a reach (Merritt and Poff 2010). Younger individuals (<two years) are less informative because they may be abundant annually but mortality is often quite high on most years (e.g., seedlings are not a good indicator of successful recruitment; Cooper and others 1999). Care should be taken to distinguish between saplings resulting from root sprouting and those germinated from seed; this should be recorded when distinguishable.

Distance along the transect where the regenerate occurs should be recorded for each individual intersecting the transect. This will provide a count of regenerates by fluvial feature for the entire reach and a calculation of number of saplings per meter.

Size-class structure

Within plots, the number of small (<2 cm diameter at breast height [DBH]) tree stems should be counted and all tree stems ≥2 cm DBH should be measured. It should be noted if tree stems occur in a cluster (e.g., they are sprouting from a larger tree). If a tree occurs near the edge of the plot, it should be measured if >50 percent of the tree stem is within the plot boundary. If stem density is relatively uniform in the plot and of high density, stems can be subsampled by counting the number of stems in a smaller measured area. If a tree or shrub occurs near the edge of the plot, it should be measured if >50 percent of the shrub or tree stems are inside the plot boundary. Histograms of size-class structure distributions may be constructed from gathered data and comparisons of distributions, central tendency (e.g., mean or median), variation (e.g., standard deviation and coefficient of variation) can be made between treatments, effected and unaffected reaches, etc. Furthermore, basal area can be calculated and summarized by species.

Condition

Shrub and tree health can be assessed visually through a simple evaluation of leaf condition (Cooper and others 2003b; Chapter 6: Case Study II). Wilting from prolonged water stress can result in leaf discoloration and partial or complete leaf death. Record the collective status of the canopy of shrubs and trees by species within the shrub-tree plots using the following categories:

- *critically stressed* = major leaf death and/or branch dieback (>50 percent of canopy volume affected);
- *significantly stressed* = prominent leaf death and/or branch dieback (20 to 50 percent of canopy volume affected);
- *stressed* = minimal leaf death and/or branch dieback (<20 percent of canopy volume affected);
- *normal* = little or no sign of leaf water stress/no water stress-related leaf death;
- *vigorous* = no sign of leaf water stress/very healthy looking canopy.

Using this ordinal scale, frequency of categories may be statistically compared between sites or reaches. Crown dieback has also been associated with increased risk of mortality in riparian trees (Scott and others 1999). Percent of potential canopy can be used to assess damage caused by water stress associated with leaf death and abscission, water stress and cavitation, and branch dieback (Scott and others 1999). Potential canopy should be estimated as a visual determination of percentage of live canopy relative to potential crown volume (i.e., extent of all branches; Scott and others 1999) for all woody species (Figure 3-13).

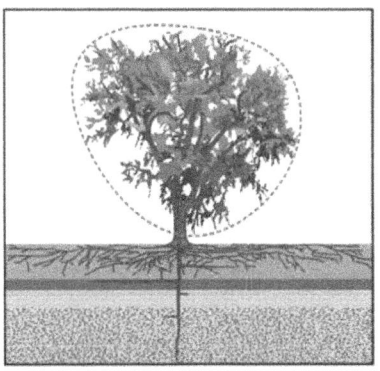

95% Potential canopy

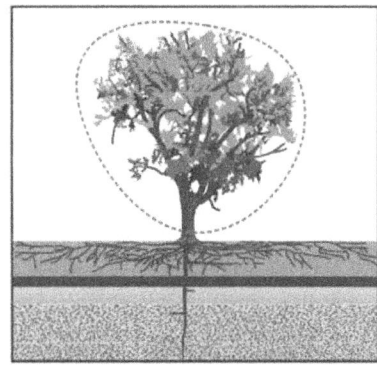

75% Potential canopy

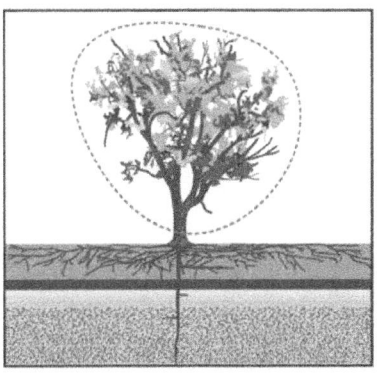

55% Potential canopy

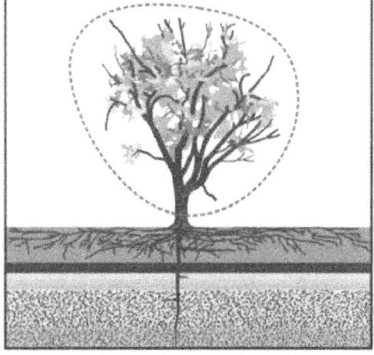

35% Potential canopy

Figure 3-13. Visual estimation of percent live canopy as a measure of condition. The observer visualizes a full canopy and then estimates the percentage of that maximum area that is occupied by canopy (to the nearest 5%).

Chapter 4: Measurement of Surface Water and Groundwater Levels

Many wetland and riparian areas are supported by both surface and groundwater, which should be considered a single resource (Winter 1999). Surface water recharges groundwater in some areas and groundwater discharges to the surface in other situations. Riparian ecosystems occur along streams and are hydrologically and geomorphically driven by surface waters, which supply shallow alluvial groundwater and influence turnover between surface and groundwater (also referred to as hyporheic exchange). Thus, stream flow, stream stage, stream dynamics, overbank flooding, and groundwater recharge or bank storage are interrelated functions. The measurement of these components is key to understanding the processes supporting these systems. The hydrologic processes supporting each wetland or riparian type should be carefully considered before designing a hydrologic monitoring program.

Fens are largely groundwater driven, and the analysis of water table depth relative to the soil surface, vertical gradients of groundwater flow, and as surface water inflows and outflows (if any exist) are critical components to measure in these systems. Fens also have little mineral sediment deposition; therefore, monitoring surface sediment deposition may be critical.

Wet meadows are also largely groundwater fed but may have surface water inflows or outflows as well. The critical features to analyze are the duration of the water table near the soil surface, duration of soil drought, and mineral sediment deposition rates. Measurement of soil redox potential over the course of a season or several seasons in wet meadows (and other wetland types) may provide insight into the range of variability typical of these systems and how hydrological alteration might change such systems.

Salt flats may be supported by surface water ponding on a relatively impermeable mineral sediment layer or shallow groundwater that supports a capillary fringe that at least seasonally reaches the soil surface. Measuring surface water inflows and outflows as well as pond water depth and duration is critical for surface water-supported salt flats.

Marshes are basins filled largely by shallow surface water, or wetlands fringing larger and deeper lakes. Measuring surface water inflows, outflows, and the depth, duration, and extent of ponded water as well as the rate of groundwater recharge is critical for understanding marsh functioning and the distribution of plant species.

These sites could be instrumented to investigate the connection between surface water and groundwater. Such instruments include piezometers (measure the hydraulic head above a well opening within confining layers), groundwater monitoring wells (measure water table level of unconfined water surface at atmospheric pressure), and staff gauges (measure surface water level) (Figure 4-1). Measurement of soil moisture and/or redox potential, particularly in wetter areas, may also provide important information about these wetlands.

Spatial scale of analysis and monitoring

The spatial scale of analysis should be determined in the planning stages, before installation of any instruments, and should include the area that may provide water, sediment, or other inputs to the study site. The distribution of measuring points and monitoring instruments should be preliminarily determined from analysis of air photos.

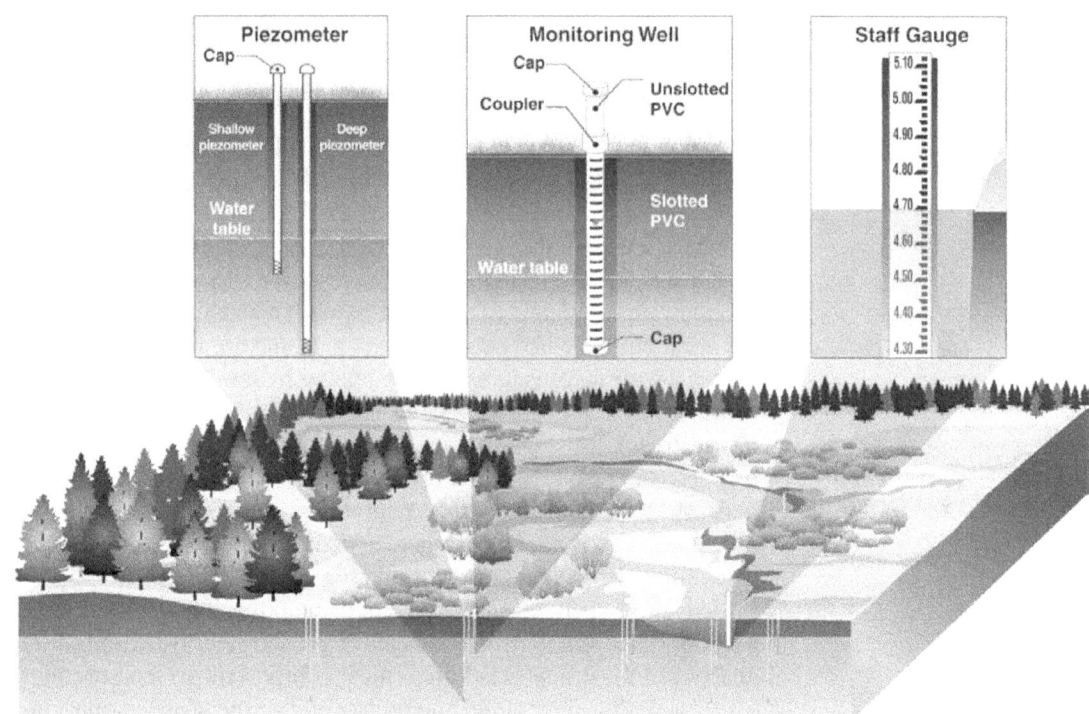

Figure 4-1. Connections between surface and groundwater may be measured using nested piezometers (which can indicate vertical flow within confined aquifers), groundwater monitoring wells (which measure water table level at atmospheric pressure), and staff gauges (which measure water surface level in a stream or water body).

Single wells only provide information about depth to water table (and its variation) below a single point. Installing grids or arrays of wells, peizometers, and staff gauges that cover the extent of the area of interest can enable the development of two- and three-dimensional water surface profiles. Surface waters should be measured at points of inflow and outflow to the study area, and water stage should be measured in areas where groundwater would be measured along transects, grids, or in plant communities that are of special interest. Groundwater should be measured in locations thought to have significant inflows to the wetland, within communities of special interest, near streams, and, in some instances, under streams.

Temporal scale of analysis and monitoring

The temporal scale of analysis should be determined in advance and be based upon the questions being addressed. For questions regarding the influence of precipitation or flow events on river stage or water tables or daily evapotranspiration-driven water table changes, data loggers associated with groundwater monitoring wells and staff gauges may provide information at the correct temporal scale. The most suitable temporal scale may vary from minutes or hours to weeks or months. Frequent data collection is needed to answer questions regarding seasonal duration of the water table within the root zone of plants. In remote areas, it may be difficult to visit the site regularly and make manual measurements. Instead, an automated measurement system that allows the collection of complete data sets can be installed with one or two visits made annually.

Approaches for measuring surface water and groundwater

Surface water

Surface water, including stream flow and pond or lake levels, is measured using staff gauges, weirs, and/or flumes. The relative or absolute level (elevation) of the water surface (also called stage) is measured using a staff gauge or staff plate, which is a measuring device anchored to the stream or lake bottom on which water level is measured. Regular measures of stage relative to the stream bottom, pond bottom, some datum below the ground surface (e.g., to keep all values positive), or a permanent reference benchmark can be made by hand and recorded in a field book. Water depth, stage, or elevation can be plotted using a simple line graph to show the water height in meters or feet above the datum, pond, or stream bottom (Figure 4-2). Regular measures (daily, weekly, or biweekly) are best for understanding the overall annual change in stage. For snowmelt-driven streams, there may be considerable daily stage change in spring and early summer due to the diurnal pattern of snowmelt. Measurements should be made at (or throughout) the same time of day, so that relative seasonal changes can be addressed. Manual measures would not provide a measure of instantaneous daily peak stage because the exact timing of peak flow and stage is unknown. In addition, some streams and ponds have stage rises during the summer that are driven by rain events, and manual measures will likely miss the exact timing and total stage change produced by these events.

If discharge can be measured simultaneously with stage, a rating curve relating water stage to discharge can be constructed (Figure 4-3A). This rating curve can then be used to calculate flow from stage height or vice versa (Figure 4-3B).

Continuous data on river or pond stage provide much more information on the maximum peak river stage, the duration of peak stage, whether multiple peaks occur, and the rate of stage change. However, since most streamflow gauge data are published as discharge, stage data may be absent or not specific to a study area. Most U.S. Geological Survey (USGS) and state gauge sites measure stage and use a rating curve constructed from a subset of field-measured stage and discharge to estimate stream discharge. During 2000, the Tuolumne River (Figure 4-4) had three major periods of high flow, one in mid-May, a second in early June, and a third in late June.

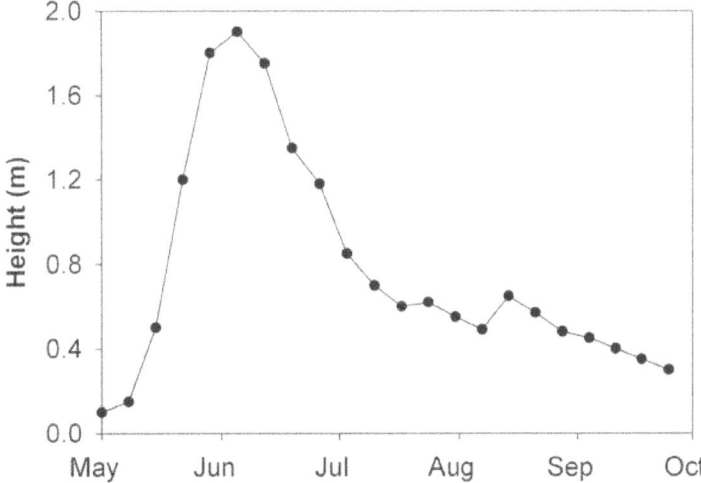

Figure 4-2. Stage (height) of water column over time based upon a series of manual measures.

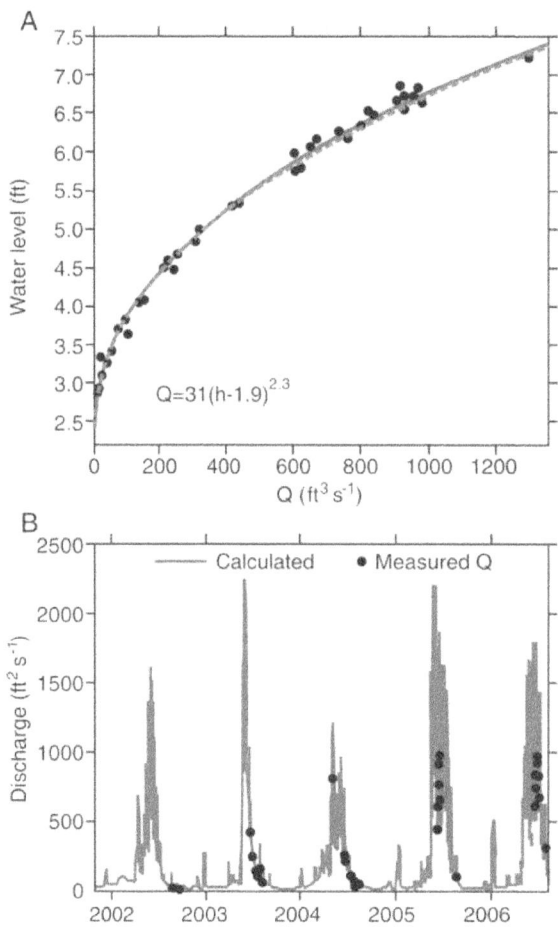

Figure 4-3. Rating curve for Tuolumne River at Highway 120 in Yosemite National Park, California, and stream flow calculated using river stage for water years 2002 to 2006. Portions of the discharge lines in the rating curve (A) indicate results from least squares fit alternating between stage or discharge as the predictor. Both yield the same equation: $Q = 31(h-1.9)^{2.3}$, where h is the staff gage reading in feet. Solid black dots on the hydrograph (B) indicate measured discharge used to construct and validate the rating curve.

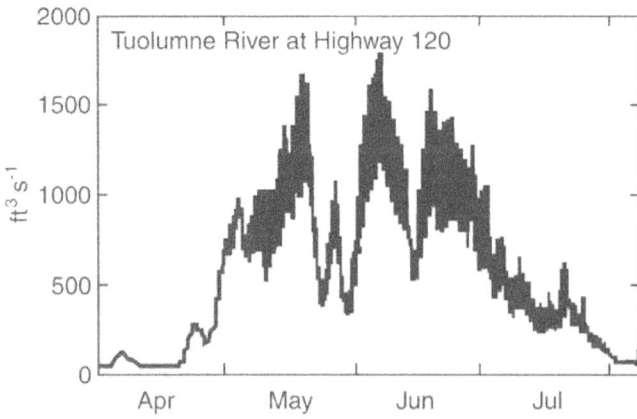

Figure 4-4. Continuous discharge for the Tuolumne River near Highway 120 in Yosemite National Park, California, from April 2000 through early August 2000. Streamflow peaked in early June followed by a cold-spell and then returned to high flow in late June, which was punctuated by afternoon thundershowers that caused small stage rises.

USDA Forest Service Gen. Tech. Rep. RMRS-GTR-282. 2012

Groundwater

Fens, wet meadows, and some salt flats are groundwater-dependent wetlands, with plants deriving most of their water from a shallow water table. Understanding the sources, flow paths, seasonal dynamics, and interactions with the soil surface is critical for managing the hydrologic driver of these ecosystems. Groundwater can flow from bedrock or from unconsolidated aquifers, or it can be recharged by surface waters such as streams and lakes. A single wetland may have more than one groundwater source. For example, different parts of a single wetland complex may be supported by a bedrock aquifer and groundwater associated with a glacial moraine, each source having distinct chemical composition and seasonality.

As previously mentioned, two primary types of instruments are used to measure groundwater in wetlands: water table monitoring wells and piezometers. Water table monitoring wells are used to measure the water table, which is unconstrained by confining sediment layers (e.g., open to the atmosphere) and is in equilibrium with atmospheric pressure. A piezometer is used to measure the pressure head in a soil or bedrock layer or at a particular depth below the water table. Other techniques for measuring groundwater are available (e.g., ground penetrating radar) but are more costly and require expensive equipment (McClymont and others 2011).

Installation of instruments

A water table monitoring well can be installed using a hand auger, a hand-held or vehicle mounted machine auger, or a backhoe (Figure 4-5). In many wetland settings, especially in peatlands and wetlands with sand/silt substrates, an option for installing water table wells or piezometers is to drive them.

Figure 4-5. Hand augers can be used to bore holes for installing a groundwater monitoring well. Mounted power augers may be used in cobble or hardpan substrate.

USDA Forest Service Gen. Tech. Rep. RMRS-GTR-282. 2012

51

Driven wells have the advantage of not requiring augering. Driven wells can, however, be problematic in areas with substantial clay content in the substrate or a hardpan. The hole or pit must not penetrate any confining layers and should be deep enough that the water table can be measured in any season and in any year. A hand-augered well could consist of a 2.5 to 10 cm (1 to 4 inches) diameter bore hole, and the layers of material that are bored through should be logged. A hand driven well casing can have its slots or holes filled with clay or other particles and the well rendered inoperable. The water table should be encountered when augering this hole, although in very fine grained sediments with very low porosity, the water table may not be apparent. Iron and manganese mottles or streaks (orange or black spots that can range from a few millimeters to a centimeter in width) may occur near the top of the seasonal water table.

Wells should be installed during the dry season, if possible, when the seasonal water table is deepest. It also can be difficult to remove sediment from a hand-augered bore hole when augering below the water table because the saturated sediment is likely to flow from the auger head. The well should extend below the water table allowing its measurement in the driest season and in a dry year.

A section of PVC pipe should be placed in the bore hole. The pipe should have machine slots or hand installed slots sawn into the pipe using a hacksaw, or it should have holes created with a drill. The slots should extend from the bottom of the casing to just below the ground surface. The holes and saw cuts should be as thin as possible so that sediment from the bore hole will not enter and fill the well casing. A filter fabric may be wrapped around the PVC to exclude sediment, but often is not needed. The bore hole around the casing can be filled to near the ground surface with clean sand or gravel of slightly larger diameter than the slots or holes. In many instances, the hole can be simply backfilled with native soils. The bore hole should be bailed using a commercial well bailer until fresh, clean groundwater fills the hole. This is especially needed when the water is muddy or contains shreds of partly decomposed peat. A bottom cap should be installed prior to placement and have a hole drilled through it to allow water to freely drain from the PVC if the water table drops below the bottom of the slots. A top cap is needed to keep rain and debris from the well casing. A monitoring well can also be used as a staff gauge to measure surface water height. In the case where the well would be used both as a groundwater and surface water monitoring device, slots or holes could be placed above the ground surface. However, in some cases, it is informative not to slot monitoring wells above the ground surface and to place clay, such as bentonite, or concrete in the top several inches of the bore hole so that surface water does not enter the sand or gravel pack around the well casing.

A piezometer is a solid section of small-diameter PVC pipe that is open only on the ends. In relatively soft soil such as peat or in moist to wet, fine-grained sediment, the PVC can be pushed to the correct depth. The bottom depth of the piezometer should be determined from the logged borings of the monitoring well. A piezometer can be installed into any soil layer for which information on pressure head is desired. A simple piezometer can be made from a section of small-diameter (e.g., 1.3 cm inside diameter) PVC pipe that is long enough to reach the depth to be measured, plus a suitable height of pipe to extend above the ground surface, typically at least 20 cm (Figure 4-6). A section of solid steel or aluminum rod (an electrical grounding rod that is copper coated steel can also be used and can be easily purchased in large hardware stores) longer than the PVC should be inserted into the PVC so that the bottom end sticks out 1 to 3 cm. The rod is held in place by locking pliers where the PVC pipe contacts the rod. The rod is held vertically with the locking pliers near the top, and the PVC assembly can be pushed to the desired soil depth.

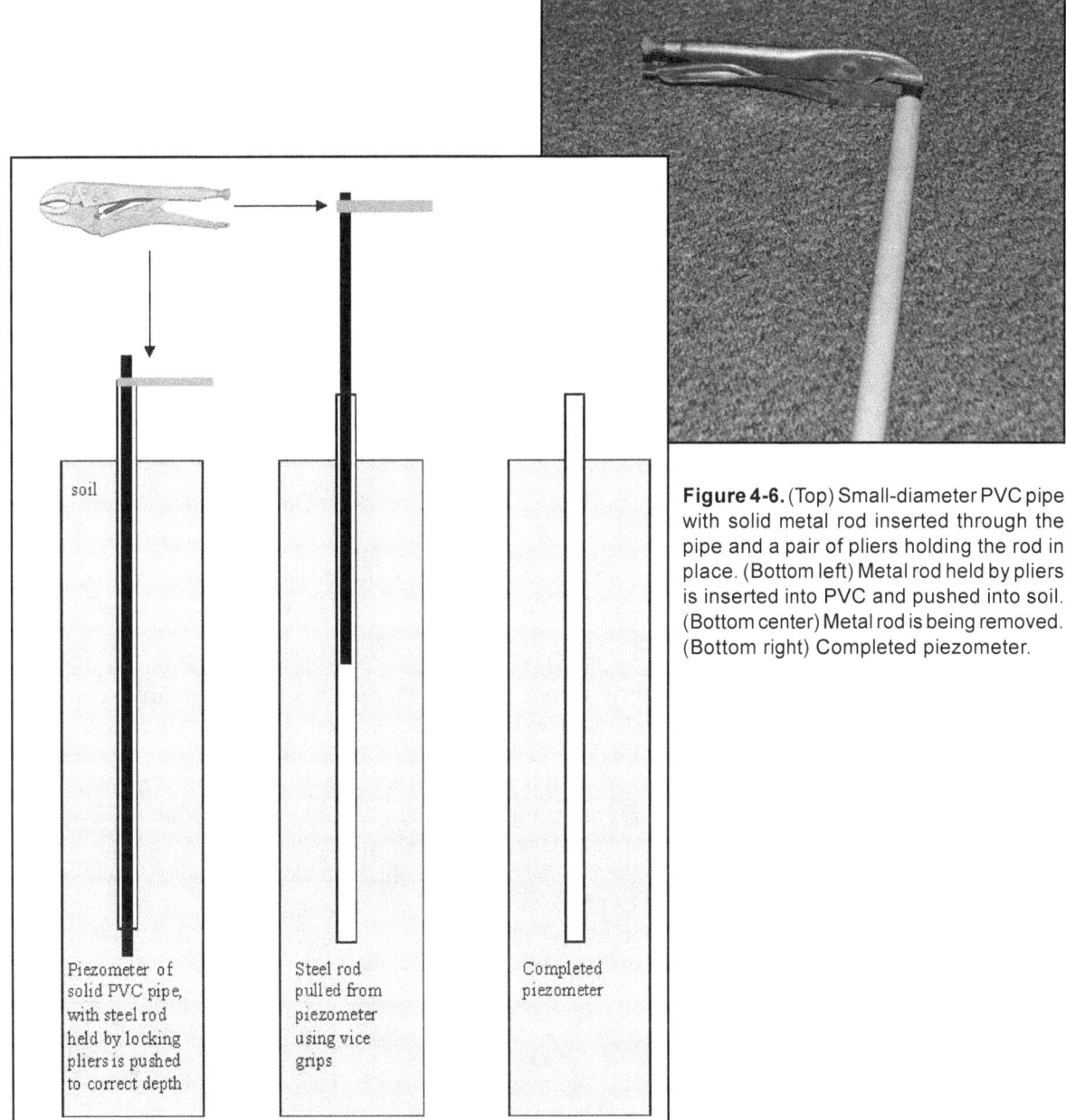

soil

Piezometer of
solid PVC pipe,
with steel rod
held by locking
pliers is pushed
to correct depth

Steel rod
pulled from
piezometer
using vice
grips

Completed
piezometer

Figure 4-6. (Top) Small-diameter PVC pipe with solid metal rod inserted through the pipe and a pair of pliers holding the rod in place. (Bottom left) Metal rod held by pliers is inserted into PVC and pushed into soil. (Bottom center) Metal rod is being removed. (Bottom right) Completed piezometer.

The pliers are used to pull the rod out of the PVC once it is inserted to the desired depth. Piezometers of this design are typically placed at several depths relative to the monitoring well to create a piezometer "nest" (Figure 4-7). Care should be taken not to auger or drive through confining sediment layers that separate aquifers. Connecting otherwise isolated aquifers may lead to the collection of data that is erroneous or difficult to interpret.

In harder material, piezometers can be constructed using augered holes. The hole should terminate at the depth the piezometer end will be placed. The PVC pipe can be slotted through the bottom 10 to 20 cm of pipe and a solid end cap should be installed. The slotting should match the soil layers that you wish to monitor. The PVC should be placed into the hole, and coarse sand should be used to fill the bore hole to the top of the slotted PVC. A layer of bentonite 20 cm or more thick should then be put over the sand, tamped in place, and wetted to encourage the clay to swell and seal the bore hole.

Figure 4-7. A well nest consisting of one water table well (right) and two piezometers.

The remaining hole can be filled with sediment removed during the boring, coarse sand, or other material. The upper part of the bore hole fill may be capped with bentonite or cement, as previously discussed.

Where soil material is too coarse or dry to auger through cobble, a drive point well may be installed (Figure 4-8A and B). Each has a cast iron drive point (red tip in Figure 4-8A) and a stainless steel slotted section (silver) with a threaded end. Threaded couplers are used to connect sections of unslotted steel pipe. A threaded drive cap is installed onto the steel pipe, and a fence post pounder is used to install the well to the desired depth (Figure 4-8B). Care must be taken to lubricate the coupler, and, using pipe wrenches, thread the coupler tightly onto the drive point and the steel pipes. If the coupler is not completely threaded on, the pounding will strip the threads. It is even more critical to get the drive cap lubricated and threaded completely onto the threaded end of the steel pipe. This facilitates driving and removing the drive cap to install an additional coupler and section of steel pipe.

Figure 4-8. Steel drive points can be pounded into the ground to create water table monitoring wells or piezometers.

Measurement

Staff gauges can be read directly from the numbers on the staff face or as distance from the top of the staff gauge to the water surface measured with a ruler. Monitoring wells can be measured with any commercial measuring tape (Figure 4-9). Weighted chalk lines and stereo wire connected to a weight and an ohm meter may also be used. The tape should be used to measure the distance from the top of the well casing to the water surface. This measurement must be corrected for the well casing height above the ground, which is called the "stick up" height of the pipe. The stick up height should be subtracted from the total depth to water table from the casing top to determine the depth to the water surface below the ground surface. Piezometers should be similarly measured. It is critical to get accurate stick up heights for all water table wells and piezometers.

Groundwater chemistry can be useful for diagnosing one or more sources of waters. Water within monitoring wells can be measured directly after first bailing the well casings out at least twice to allow fresh groundwater to enter the pipe. Water in streams, ponds, or monitoring wells can be directly measured using an electrical conductance meter, or water can be collected for analysis of cations, anions, or nutrients.

Figure 4-9. Manual measure of water table depth in monitoring well using a tape measure.

Redox potential

Water is denser than air, and when soils flood or a water table rises to saturate soils, water forces air out of the soil interstices. If the soils are warm enough for biological activity, then bacteria and plant roots can remove the remaining dissolved or trapped oxygen (O_2) in the soil. At this point, the soil is anaerobic or anoxic, meaning that it lacks free O_2. Plants that must obtain O_2 for root metabolism directly from the soil they are rooted in will drown if the soils remain anoxic for more than a couple of weeks during the growing season. Once free O_2 in the soil is depleted, bacteria that can use molecules other than oxygen as their terminal electron receptor for respiration processes become active. Since electrons have a negative charge, when an electron is added to a molecule such as ferric iron (Fe^{3+}) or manganic manganese (Mn^{4+}), it reduces (makes more negative) the molecule's electrical charge, producing ferrous iron (Fe^{2+}) or manganous manganese (Mn^{2+}). This process is called reduction and such soils are considered to be "reducing". A series of biogeochemical changes occur in soils as they become increasingly reduced, with oxidized forms of nitrate (NO_3^-), Mn^{4+}, Fe^{3+}, sulfate (SO_4^-), carbon dioxide (CO_2), and hydrogen (H) being reduced. Reduced forms of Mn and Fe are soluble in water, and plants that uptake them can suffer heavy metal poisoning. Nitrogen reduction removes NO_3 from the soils by producing gaseous N. Hydrogen sulfide (H_2S), formed by the reduction of SO_4, is toxic to plant roots. Carbon dioxide reduction forms the important greenhouse gas methane (CH_4).

Redox potential is a measure of the oxidation state of various reduction couples in soils. The soil oxidation-reduction potential can be measured with a millivolt meter using a pure platinum tipped electrode coupled with a reference electrode to complete the redox circuit. The platinum must be pure (>99 percent platinum) wire and thick enough that it can be inserted into the soil, such as 18-gauge wire. An approximately 1 cm long piece of platinum must be fused to copper or brass wire or rod without the introduction of an additional metal such as solder. If copper is used, it should be pure. The copper or brass is heated with a torch until it just melts. While heating the copper, the 1 cm long piece of platinum wire is held with a pair of needle nose pliers. When the copper melts, the heat is removed and the platinum wire is touched to the copper simultaneously, allowing the metals to fuse. The copper and fused junction must then be sealed using waterproof heat shrink tubing or another substance. The junction between the platinum and copper must be completely sealed because any copper exposed to the soil environment will foul the circuit. Detailed instructions for making platinum tipped electrodes are presented in Wafer and others (2004) and in Mitch and Gosselink (2000).

The platinum electrode is inserted into the soil to the depth desired for measurement. A calomel (Ag/Cl) reference electrode is also inserted into the soil, and both electrodes are attached to a millivolt (Mv) meter, which measures the electron flow in Mv. Most high-quality pH meters have a Mv mode. The reference electrode is attached to the reference jack, and the platinum electrode must be fitted with a Bayonet Neill-Concilman (BNC) end to attach to the pH electrode jack. The Mv readings must be corrected for the reference electrode by adding +244 Mv to the reading. In addition, pH can influence redox readings, and the raw data must be corrected by -60 Mv per pH unit greater than or less than 7.0.

It is most reasonable to use redox potential to provide an indication of the general oxidizing or reducing condition of the soil. Broad distinctions such as oxic (free O_2 is present in the soil, and redox potential >+400 Mv), moderately reducing (+100 to +400 Mv), reduced (+100 to -100 Mv), and highly reduced (-100 to -300 Mv) are suggested (Bohn 1971).

Soil redox potential can be highly variable depending upon the microhabitat that the platinum electrode contacts in the soil. Therefore, it is suggested that at least three electrodes be installed at each soil depth of interest. In practice, it's best to install the electrodes and leave them in place for the entire study period to disturb the soil as little as possible. Redox potential measures are most reliable in saturated and reduced soils. Measures should be performed often enough to allow the seasonal patterns of redox potential to be revealed. The measures are most useful if coupled with a groundwater monitoring well that is equipped with a pressure transducer in order to record daily water levels at the study site. This is particularly important in sites with highly variable hydrologic regimes. For example, a researcher may visit a field site that has saturated soils on the day of the visit but he/she may measure oxic soils. This would make sense if the site had recently become saturated and had insufficient time for reducing conditions to develop. However, the opposite could also happen. A site could appear dry at the surface but because of previous hydrologic conditions, it could have been saturated for many weeks and the soils could be reducing as it may take days to weeks for air to reach the soil depths being measured (particularly in fine-textured soils). Thus, data on recent hydrologic patterns and processes are critical for interpreting soil redox potential measurements.

Soil redox potential is as important to plant species distribution as is water table depth or its duration. This is because highly reducing conditions produce inhospitable environments that relatively few species can survive.

Data presentation

Water table wells and piezometers

Depth to water for monitoring wells can be illustrated as simple line graphs showing the trends in one or more wells relative to the ground surface (Figure 4-10A) or true elevation of the water level (Figure 4-10B). This allows a direct comparison of water levels among wells. In Figure 4-10A, water levels for 15 wells are compared for two summers for Cottongrass Fen in Colorado's San Juan Mountains. The year 2003 was dry and water levels in many wells dropped well below the ground surface, while 2004 was a near average snow year and water levels for most wells remained near the soil surface (other than well CW11). Figure 4-10B compares a staff gauge in the Merced River (X16) and 10 monitoring wells (33-77) on the Yosemite Valley floor in Yosemite National Park, California, using true elevation of the groundwater and surface water.

Water levels in piezometers should be compared to each other and to the adjacent monitoring well (Figure 4-11). A piezometer with head higher than the water table indicates an upward hydrologic gradient in the sediment layer being monitored by the piezometer. A piezometer with head lower than groundwater in the well indicates a downward gradient, while a similar head (relative height of water surface) and water table indicate minimal vertical flow and suggest that groundwater flow is roughly horizontal. In Figure 4-11, SpW1 has downward flow in the piezometer with a terminus at 147 cm below the soil surface, while the other piezometers have heads above the water table. All piezometers at site SpW2 have upward flow in all three years, while at StW2 flow is primarily horizontal.

Cross sections and profiles

Cross sections and longitudinal profiles that illustrate both the land surface and the water table and/or piezometric head are important tools for showing the relationship of the water table to the land surface over large areas, and they can be used to infer flow direction.

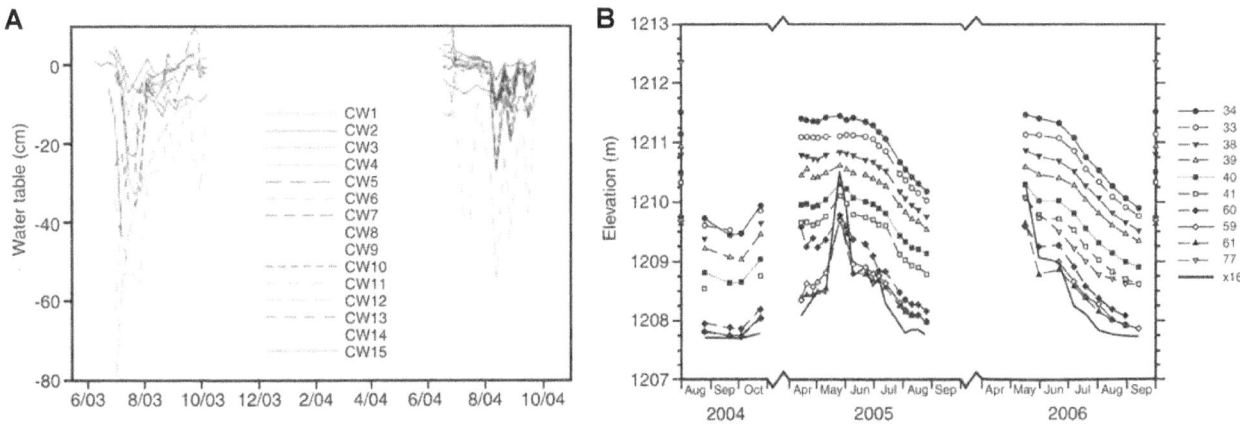

Figure 4-10. (A) Water table presented as depth below ground surface and (B) as elevation for multiple wells.

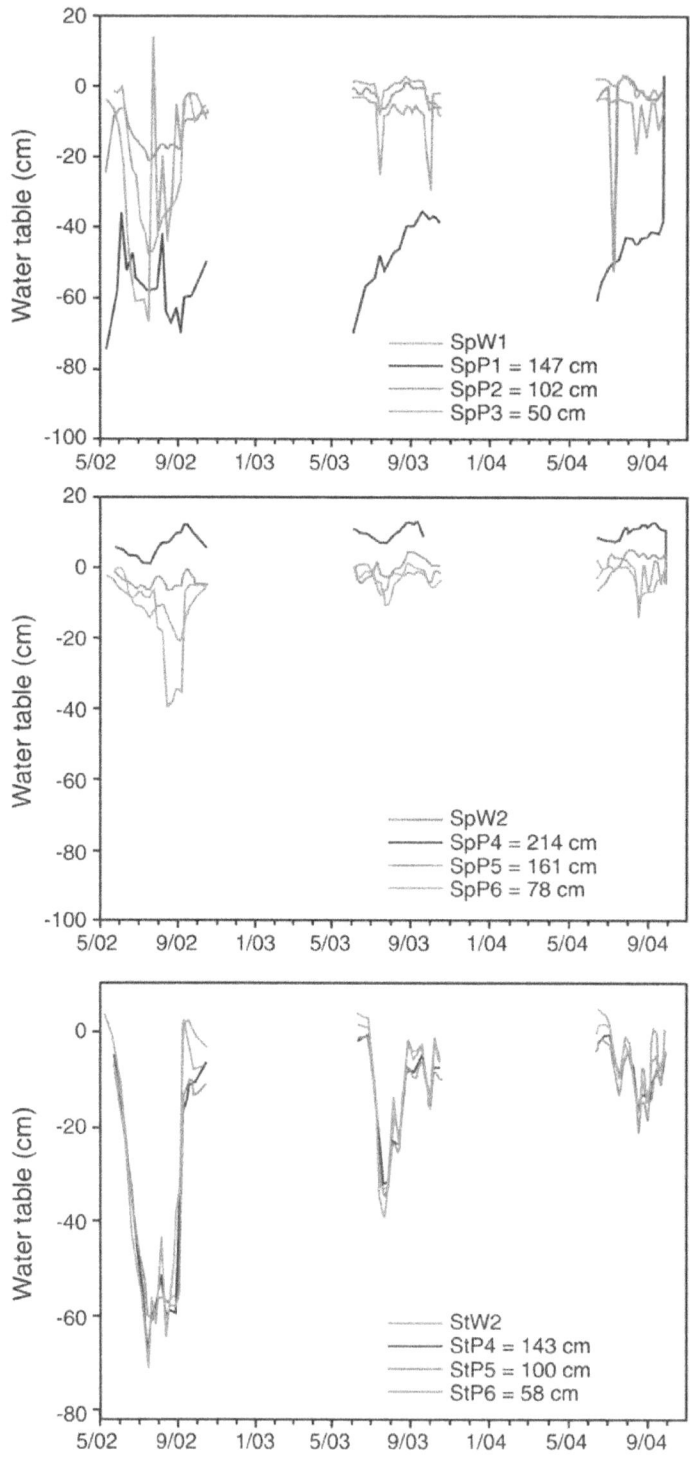

Figure 4-11. Nests of one water table monitoring wells (W) and three piezometers (P) located adjacent to each well. SpW1 is Spruce fen Well 1, SpW2 is Spruce fen Well 2, and StW2 is String fen Well 2 in the Prospect Basin area of the San Juan Mountains near Telluride, Colorado. The green lines are the water table depth below the soil surface (0), and the three piezometers are used to measure the head at three depths in each location. The completion depth, in centimeters, is listed for piezometers.

Figure 4-12 is a 1200-m longitudinal profile across the western portion of Tuolumne Meadows in Yosemite National Park, California, that shows the ground surface elevation and groundwater levels for four dates during 2006. The wells near the road are in upland conifer forests (wells 62, 56, 55), while the other wells are in wet meadows. Groundwater flows from the uplands on the left toward the Tuolumne River on the right. Wells 10, 11, and 71 appear to have water levels that closely follow the river stage, indicating that these are riparian sites, while the remainder of the meadow along this transect is a groundwater-fed wet meadow.

Many mountain streamside wetlands are groundwater-fed and supply streams with water, as can be seen for Snow Spur Creek on Lizard Head Pass at the headwaters of the Dolores River in Colorado (Figure 4-13A and B; cover photo). In this area, Snow Spur Creek is a gaining stream. The San Miguel River, near Uravan, Colorado, is a losing reach, with a higher stream stage than the adjacent floodplain groundwater elevation.

Water table maps

Water table elevations for wells that are organized in a spatial grid can be used to make a two-dimensional water table map that illustrates the overall water table elevation as well as the direction and gradient of groundwater flow. The map in Figure 4-14 for Tuolumne Meadows shows groundwater flow from the south side of the valley toward the channel (gaining reach) and flow that is parallel with the Tuolumne River in the right side of the diagram. Flow is from right to left.

A water table map can also be used to show water table contours near a stream, such as for the Snake River in northwest Wyoming (Figure 4-15). This figure shows groundwater flow within a floodplain, with water having been lost from the river upstream near letter A and flowing back to the river downstream near B.

The water table elevation can also be portrayed as a color contour map, shown for Yosemite Valley on 17 May 2006 (Figure 4-16). This figure shows groundwater flow from north to south from Ahwahnee Meadow and from south to north from Stoneman Meadow. Both groundwater flow systems feed the Merced River, which is flowing from right to left.

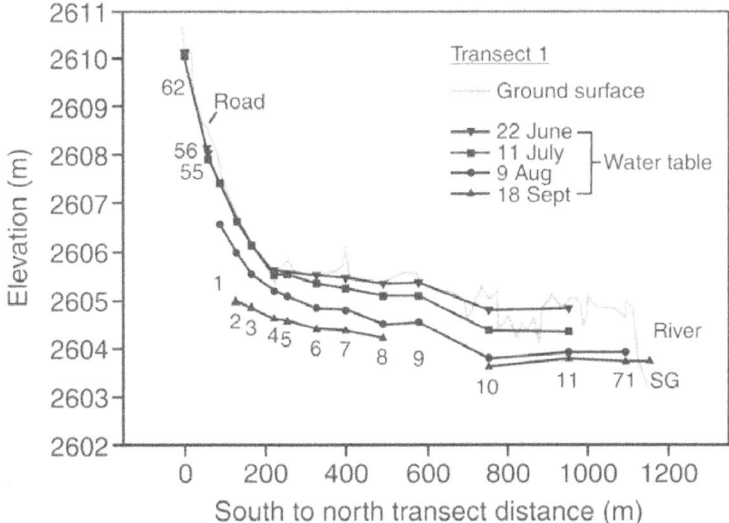

Figure 4-12. A 1200 m long transect across the western side of Tuolumne Meadows in Yosemite National Park, California. Water levels recorded in 15 monitoring wells (numbered) and 1 staff gauge (SG) are shown for four dates during 2006. The Tuolumne River is at right, and a hillslope bordering the meadow is at left. The Tioga Pass highway is shown at ROAD.

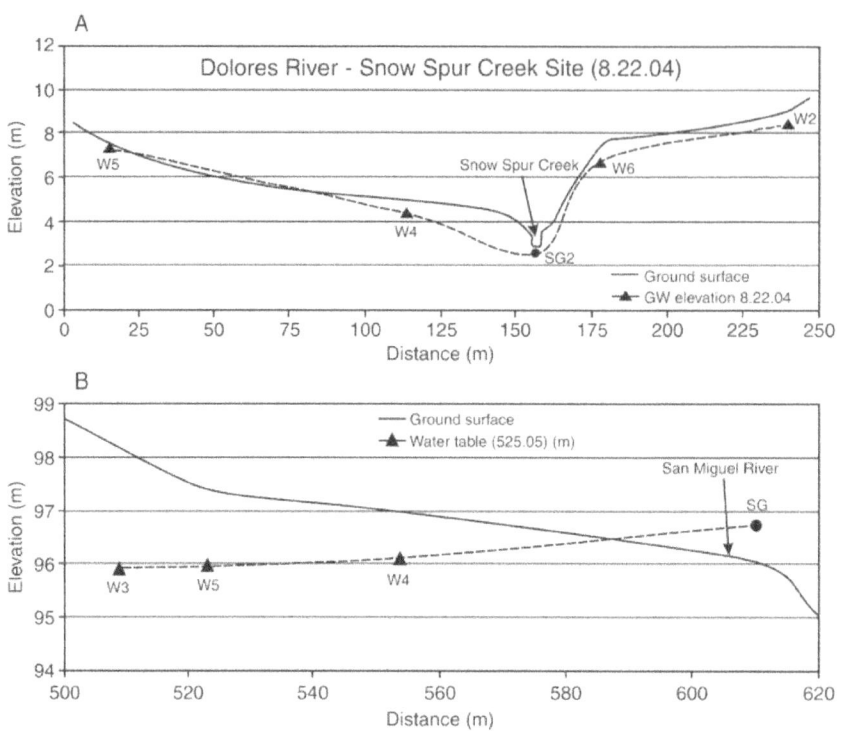

Figure 4-13. Ground surface and surface and groundwater elevations on two profiles. Snow Spur Creek (top) and San Miguel River (bottom).

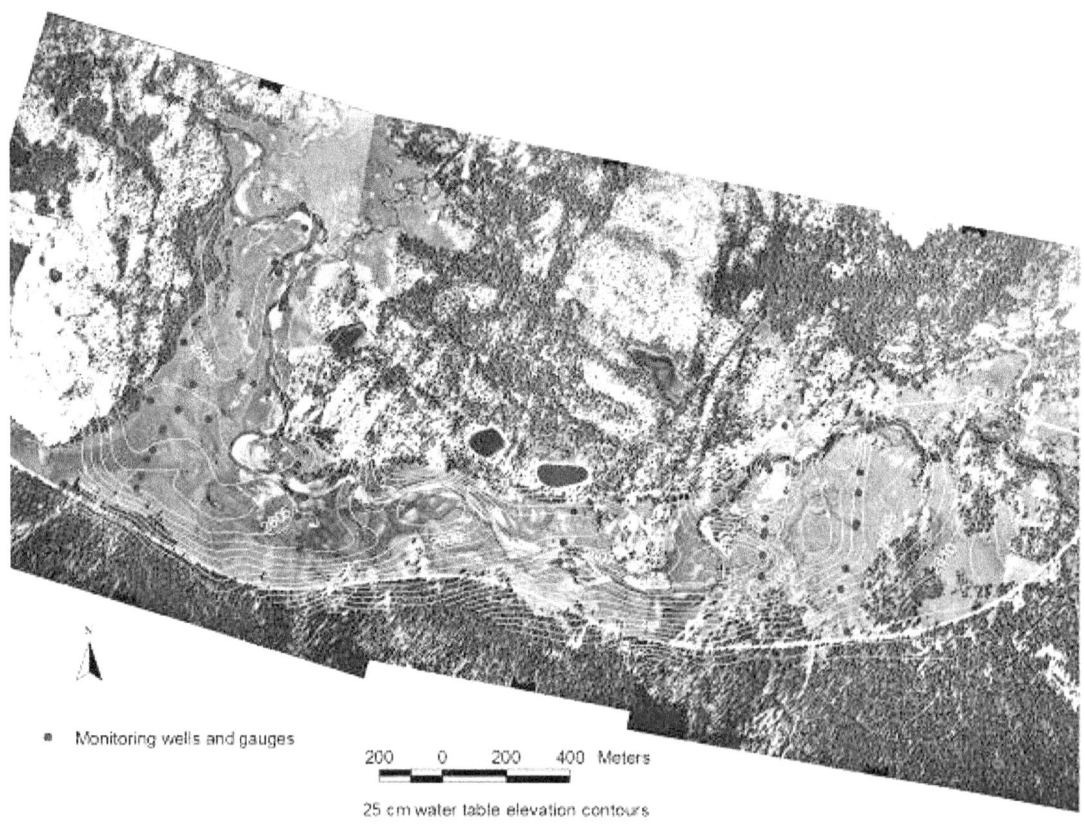

Figure 4-14. Water table contour map of Tuolumne Meadows, Yosemite National Park, California, for 18 September 2006. From Cooper and others (2007).

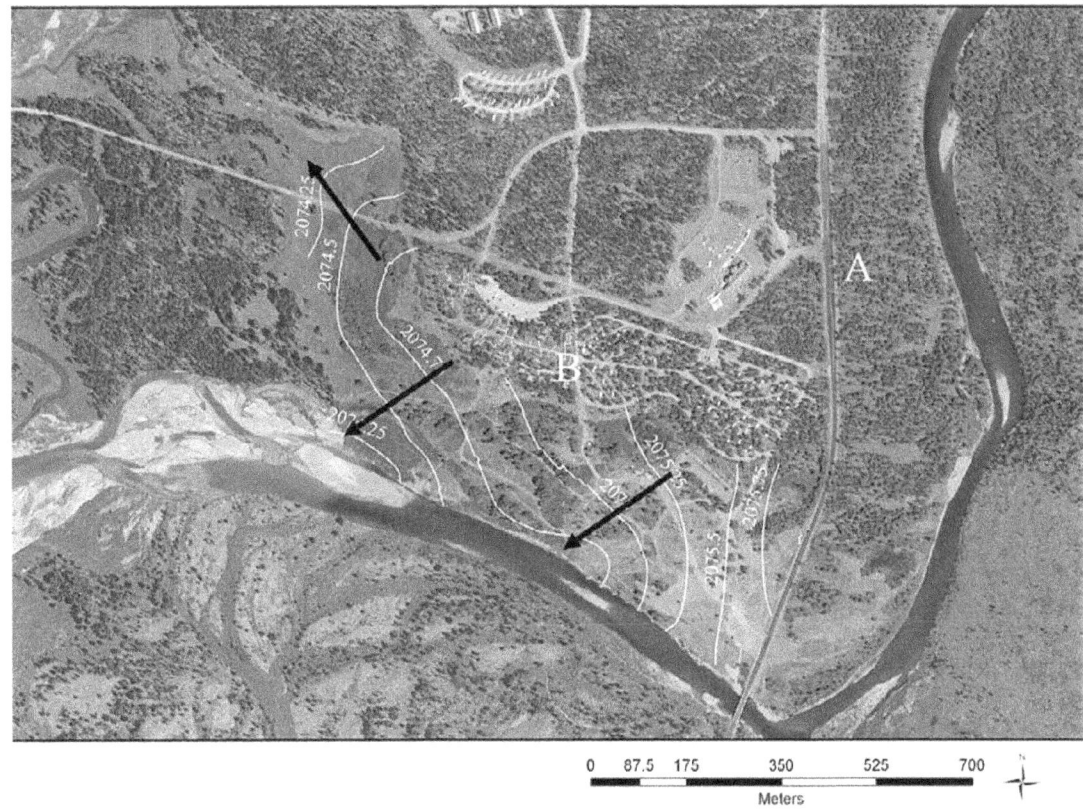

Figure 4-15. Water table contour map for the Snake River floodplain at Flagg Ranch, Wyoming, with arrows indicating the direction of flow. From Cooper and Patterson (2007).

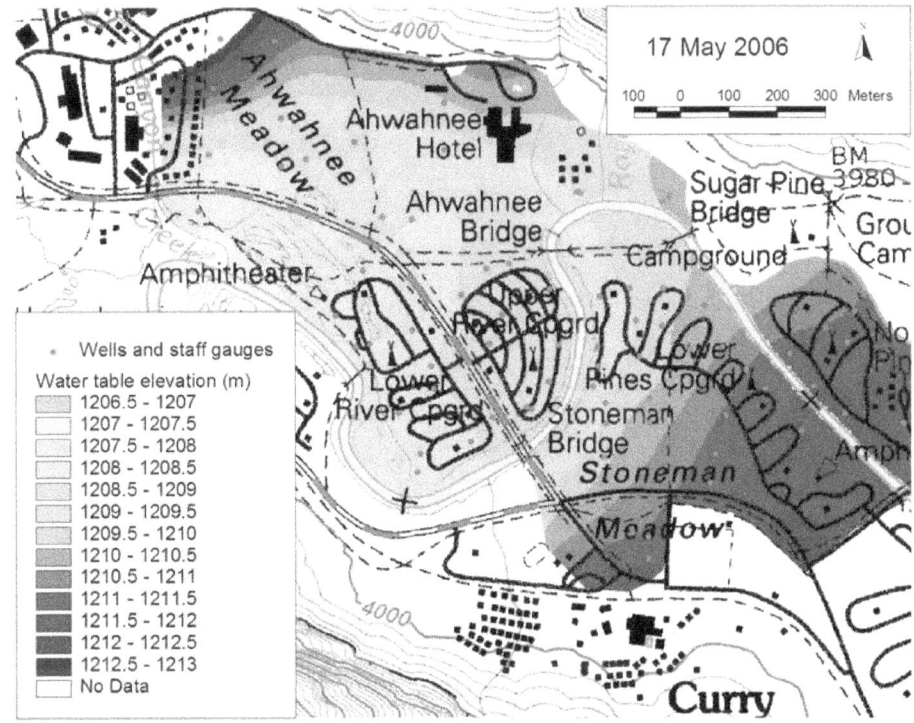

Figure 4-16. Water table contour colored map of Yosemite Valley. From Cooper and Wolf (2006).

Depth to water table map

The water table depth below the ground surface may also be presented in two-dimensional map form. This type of figure (Figure 4-17) should be presented for different seasons (for example, early summer and late summer) or for a wet versus a dry period. It can be used to highlight areas affected by ditches and the effect of restoration on the water table depth; for example, the restoration of Big Meadows fen in Rocky Mountain National Park, Colorado (Figure 4-17; top panel is before ditch restoration, bottom panel is after restoration).

Data management

The management of data is essential to storing and using data for short-term and long-term analyses. Raw and post-processed data can be stored using any spreadsheet program or database. An example of the data that should be collected in the field and used to produce a high-quality data set is shown below (Table 4-1). For each water table well or piezometer, the well number, total length of the PVC pipe, length of pipe below the ground surface, and length of pipe above the ground surface (stick up) must be accurately recorded. The elevation of the casing top should be accurately determined.

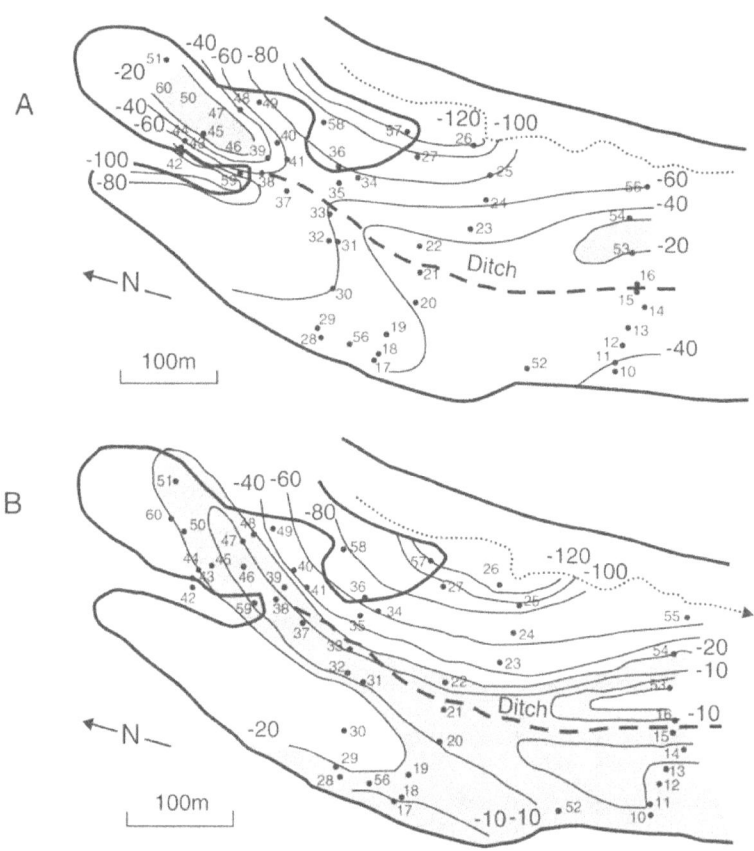

Figure 4-17. Depth to water table map for Big Meadows, Rocky Mountain National Park, Colorado. From Cooper and others (1998).

Table 4-1. Well number, total length of PVC pipe used to create the well, length of PVC belowground, length of pipe aboveground ("Stick up"), elevation of the casing top ("Case elevation"), depth to the water table from the top of casing ("07-Jul-08"), depth to the water table below ground surface ("WT depth"), and the elevation of the water table ("WT elev."). All units are meters.

Well #	Total length	Belowground	Stick up	Case elevation	7-Jul-08	WT depth	WT elev.
1	3.12	2.91	0.21	3221.65	1.58	1.37	3220.07
2	2.75	2.55	0.20	3222.05	2.20	2.00	3219.85
3	1.68	1.45	0.23	3221.90	1.55	1.32	3220.35
4	3.85	3.60	0.25	3223.20	3.22	2.97	3219.98
5	2.99	2.71	0.28	3223.20	2.26	1.98	3220.94

Raw measures of water table depth should be entered completely, under a column for the date on which the measures were made. The raw measure from the casing top includes the stick up length, which should be subtracted from the raw measure to calculate water table depth below the ground surface. The raw measure should be subtracted from casing elevation to determine elevation of the water table. Water table depth is used to create hydrographs and maps of water table depth, while water table elevation is used to create profiles and cross sections, and water table elevation maps. Each period of data collection could be entered into this spreadsheet in a column to the right of the 7 July 2008 data. The water table depth and water table elevation calculations could occur on a linked worksheet.

Topographic surveying

Accurate topographic data on the true or relative elevation of the ground surface and on the location and elevation of measuring instruments is essential for almost all hydrologic investigations. If a single monitoring well is being analyzed and depth to the water table is the only measurement needed, then accurate topographic information may not be necessary. However, to understand the relationship between two wells or sets of instruments of any kind, topographic information is needed. Topographic surveys can be conducted with a variety of instruments, including laser total stations (Figure 4-18) and GPS based surveying equipment, or a more traditional survey can be conducted that utilizes level or theodolite to measure distances and angles

Figure 4-18. Topographic surveying with full station in Death Valley National Park, California.

USDA Forest Service Gen. Tech. Rep. RMRS-GTR-282. 2012

63

between points. A topographic map can help a researcher decide where to install instruments and is critical for making water table maps. Cross sections and profiles can be generated if the distance between wells is calculated using a tape measure and if the relative height of each well is measured using a rod and level. Detailed descriptions of surveying are presented in Harrelson and others (1994), which is available at: http://www.stream.fs.fed.us/publications/PDFs/RM245E.PDF. Guidelines for surveying cross sections, measuring flow velocity, and calculating stream discharge are provided in Buchanan and Somers (1976), which is available at: http://pubs.usgs.gov/twri/twri3a8/html/pdf.html.

Hydrologic study design

The spatial distribution of instruments and the types of instruments used depend upon the landscape being analyzed, the questions being addressed, and the skills of the researcher. Measuring devices (instruments) may be aligned along transects where known gradients exist; for example, traversing an upland-wetland gradient or a stream channel-floodplain gradient (Figure 4-19). A grid is suitable where unknown hydrologic patterns and processes are being investigated and where the construction of a water table map is desirable. Hydrologic investigations may also focus on the known or observed distribution pattern of plant communities or on other ecological or hydrologic processes.

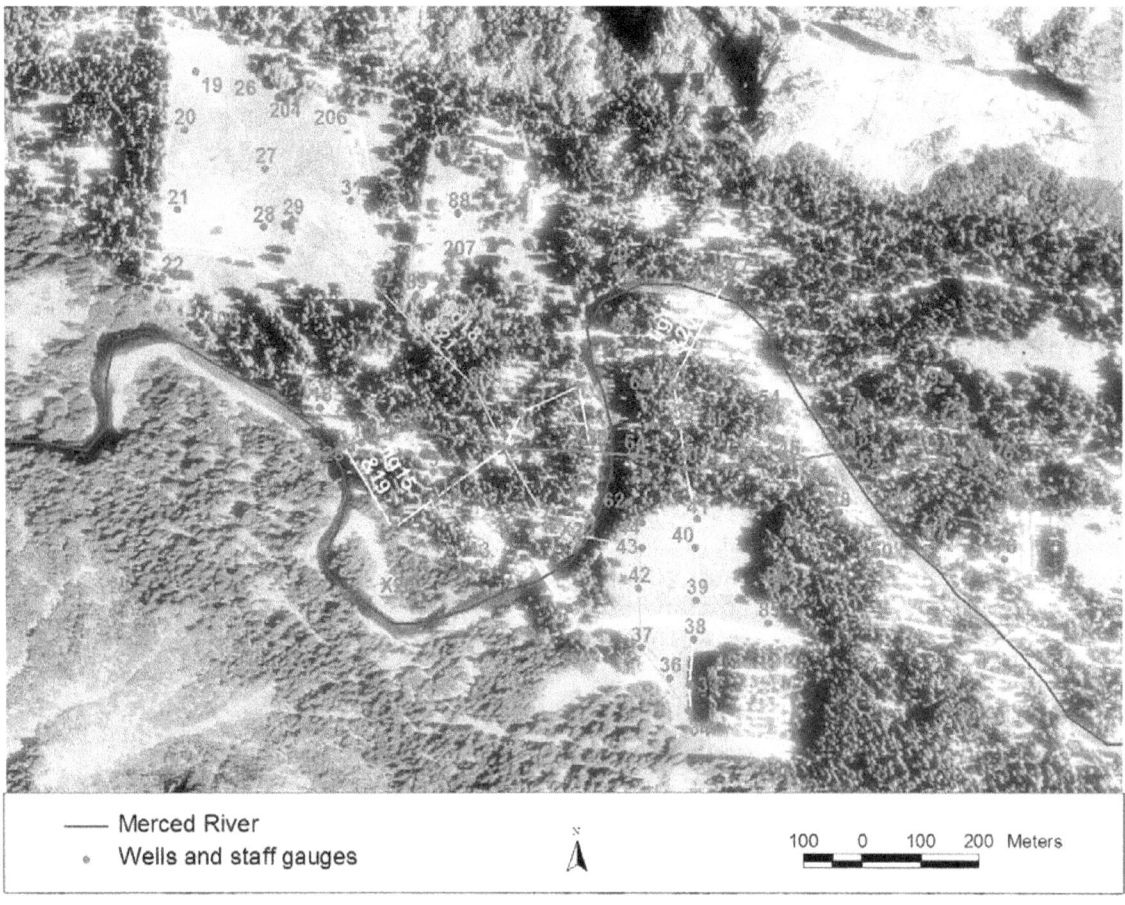

Merced River
Wells and staff gauges

100 0 100 200 Meters

Figure 4-19. Distribution of monitoring wells (red numbers) in Yosemite Valley, California. These wells are set as a series of transects that investigate the relationships of upland to lowland, river to floodplain. The transects are oriented parallel to and perpendicular to each other, which facilitated construction of water table maps. From Cooper and Wolf (2006).

This allows for the collection of data on resources of interest and for the comparison of water levels among plant community types.

Groundwater-surface water interactions

The interactions of surface water and groundwater support many key functions of streams and link ecosystems through water and nutrient exchanges. Several patterns provide strong evidence for the linkage between surface and groundwater. If stream stage is nearly identical to the pattern and rate of change of groundwater, they are most likely connected, rising and falling in tandem (Figure 4-20). This can also be shown in a profile oriented at 90 degrees to the flow direction of the river (Figure 4-21). In this example, stream stage is higher in elevation than groundwater in any monitoring well for the four measurement dates, suggesting flow from the river into the floodplain soils.

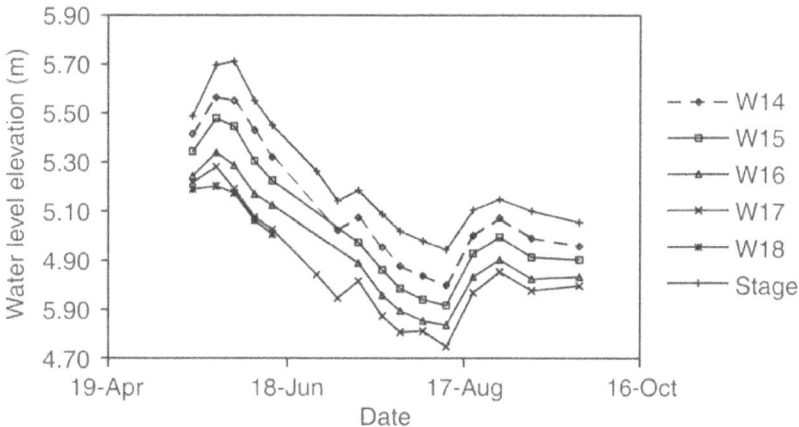

Figure 4-20. Stream stage is completely correlated with groundwater levels in five monitoring wells on the floodplain of the San Miguel River in southwestern Colorado. This suggests that stream water supplies and controls the elevation of groundwater in this location. From Cooper and Arp (1999).

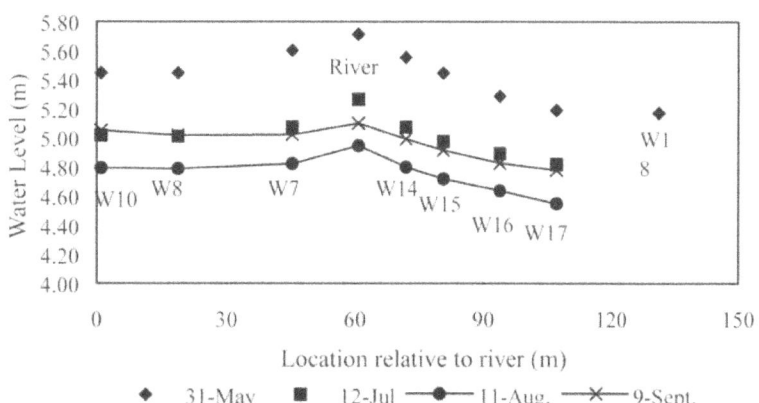

Figure 4-21. Measures of relative stream stage and groundwater levels along the San Miguel River in Colorado. Each symbol is a staff (river) or well, and their distance from the river staff is measured in meters. From Cooper and Arp (1999).

Surface water and groundwater interactions can also be shown with water table maps. This requires both staff gauges in the stream and a network of monitoring wells. All wells and staff gauges must be topographically surveyed for position and elevation. Where the flow net (water surface and water table elevations) indicates movement of water from the river to floodplain, it strongly suggests groundwater recharge by the stream (Figure 4-22). Surface water and groundwater interactions can also be investigated using geochemical comparisons of surface water and groundwater by adding tracers to surface water and other techniques.

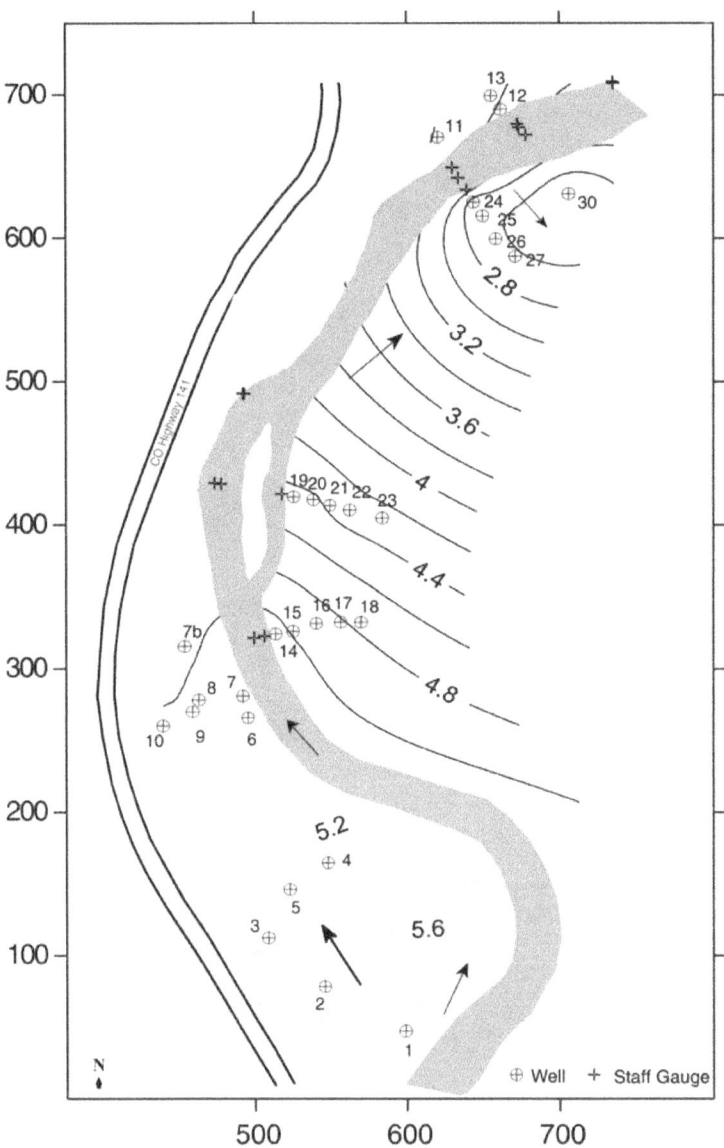

Figure 4-22. Water table map for the San Miguel River, near the Uruvan gauge, showing 0.5-m contour lines. Monitoring well and staff locations are identified. The flow net indicates strong flow from the river into the floodplain in certain areas, such as the bottom of the figure, and on river right from 150 to 300 m along the y axis. In other areas, groundwater flow parallel with the river is indicated by flow lines perpendicular to the river. From Cooper and Arp (1999).

Hydrologic impact analysis: Groundwater pumping

Groundwater is pumped for human use in many areas of the world. Water pumped from shallow depths may influence the local or regional water table elevation. A groundwater pumping well occurs in Crane Flat meadow in Yosemite National Park and its effects have been analyzed (Figure 4-23A; blue dot within the black rectangle). A grid of nested water table wells and piezometers was established across Crane Flat (Figure 4-23B). Three well/piezometer nests—A-1, A-2, and A-3—are illustrated here (Figures 4-23C and D). Figure 4-23C illustrates three water table well and piezometer nests (A-1, A-2, and A-3); the water table is indicated by the white bar with cross hatching and piezometers are indicated by the solid bars. The ground surface (top of gray shading) and the soil zones that are saturated (blue) and unsaturated gray. Figure 4-23C shows a period in early summer prior to the initiation of pumping, while Figure 4-23D shows the same site during the mid-summer pumping period. The decline of the water table (shown as the top of the blue shaded area) is apparent, as are the heads in the piezometers (shown as red triangles across each piezometer). A continuous logger in well 49 (Figure 4-23E) illustrates the rapid drawdown during July and August, with diurnal fluctuations created by pumping for approximately 12 hours per day and a rain in October that led to water table recovery.

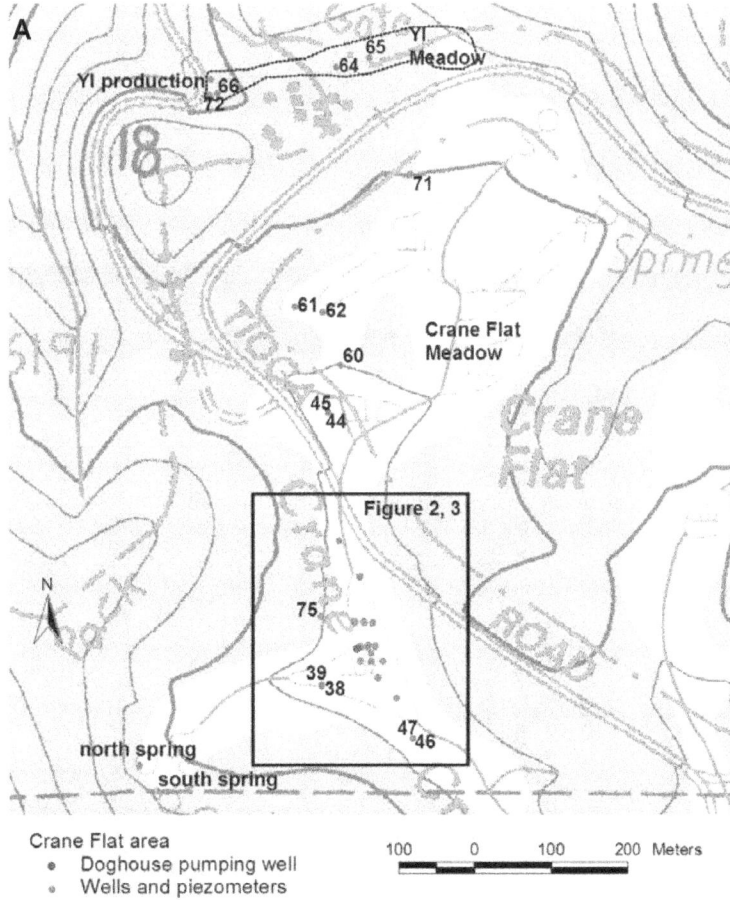

Figure 4-23. Location of monitoring wells (A), detailed view of well networks, including nests of wells and piezometers (B, C, and D) and patterns of hourly water levels as influenced by pumping. Pumping ceased in late October. Light blue area in (C) and (D) indicates groundwater level in early July (C) versus mid-September (D). From Cooper and Wolf (2007).

Figure 4-23. (Continued)

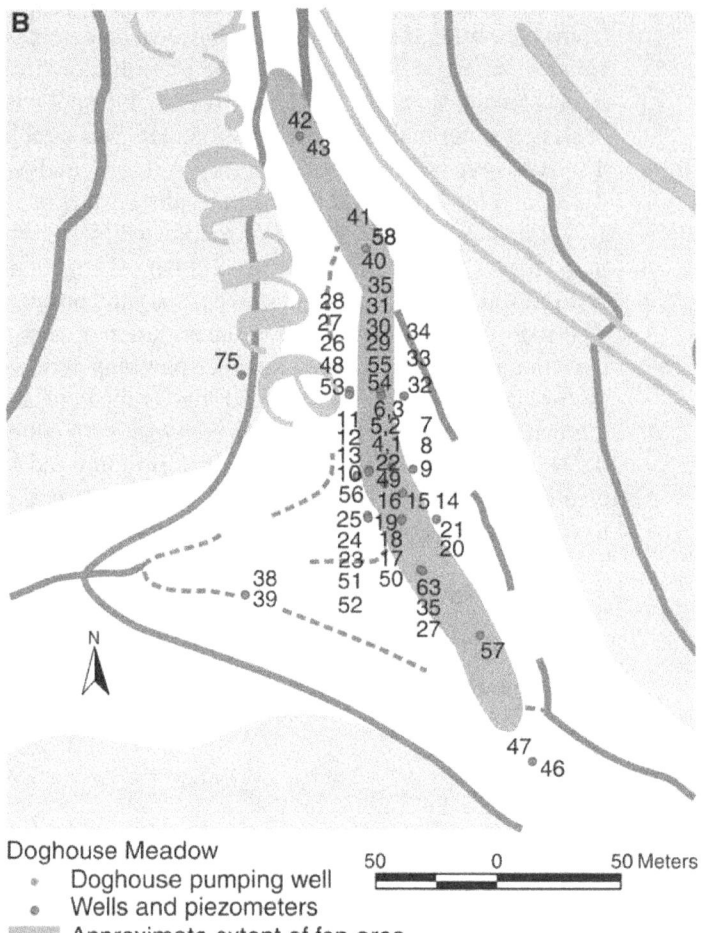

Doghouse Meadow
• Doghouse pumping well
• Wells and piezometers
▨ Approximate extent of fen area

50   0   50 Meters

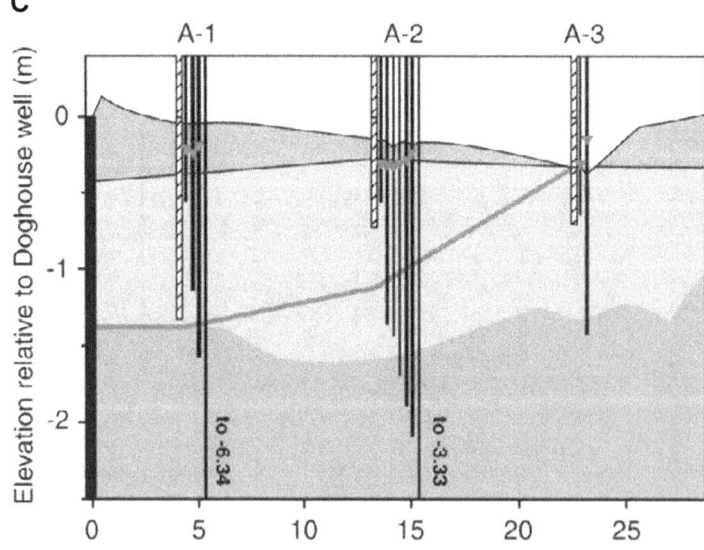

Figure 4-23. (Continued)

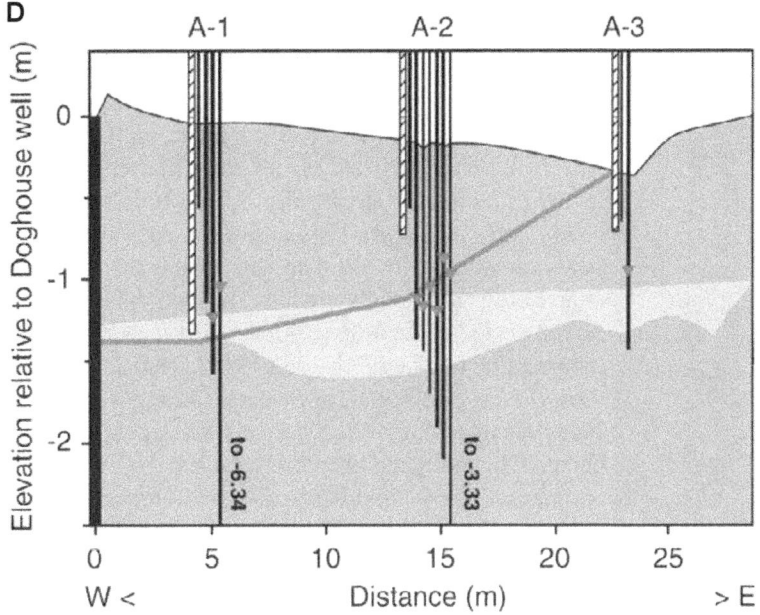

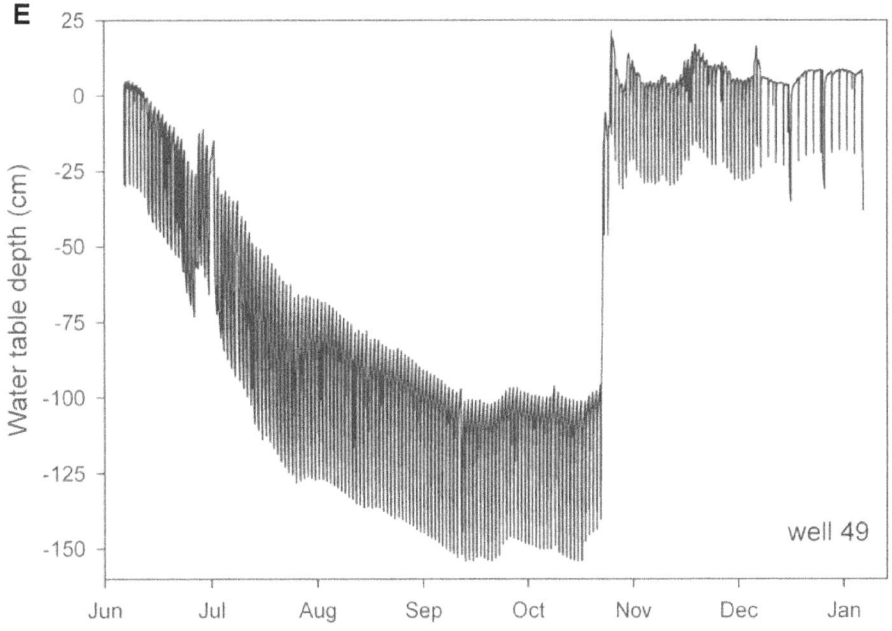

Hydrologic impact analysis: Changes in stream flow

Stream flows have been altered by human constructed dams, water diversions, and other structures. Climate change during the past 100 years has also been shown to have changed the flow of many streams (Karla and others 2008). Two methods for showing changes in stream discharge are plotting mean daily flows and annual instantaneous peak flows. Figure 4-24 plots the mean daily flow for the Green River at the Greendale in Wyoming, located just below Flaming Gorge Dam. The dam was completed in 1962 and altered stream flows starting in spring 1963. Flows below the dam were nearly curtailed during 1963 and 1964, and post-dam peak flows were reduced in most years compared with pre-dam years. Base flows have been increased, thereby reducing annual flow and stage variance. Through the early 1990s, a single seasonal snowmelt-driven period of high flow did not occur, and high and low flows occurred throughout the year due to dam operations. Five high flows occurred in the 1980s and 1990s when the reservoir pool filled following winters with very large snowpack in the headwaters. Flow management produced a seasonal single peak beginning in the 1990s to meet the needs of federally listed endangered native fishes. Figure 4-25 illustrates instantaneous annual peak flows and the striking difference between pre-and post-dam years. Most post-dam years have nearly identical peak flows due to the capacity of the electricity generating turbines to pass water at a rate of 130 m^3/s (4600 ft^3/s).

Commercial software is available for characterizing streamflow regimes and for analyzing hydrologic impacts to streams. Indicators of Hydrologic Alteration (IHA) is software that calculates 64 statistics summarizing daily average flow data either before-and-after date of impact or over time (range of variability analysis). IHA software and instructions for its use are available from: http://conserveonline.org/workspaces/iha/documents/download/view.html. Olden and Poff (2003) reviewed metrics for characterizing streamflow and linking them to biological patterns, including many of the metrics generated in IHA.

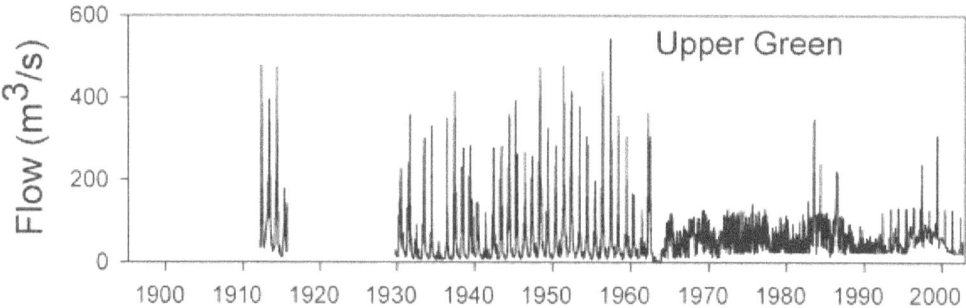

Figure 4-24. Mean daily discharge of the Green River at the Greendale (USGS gauge number 09234500) and Linwood (USGS gauge number 09225500) gauges in Utah. No data were collected prior to 1912 and between 1915 and 1929.

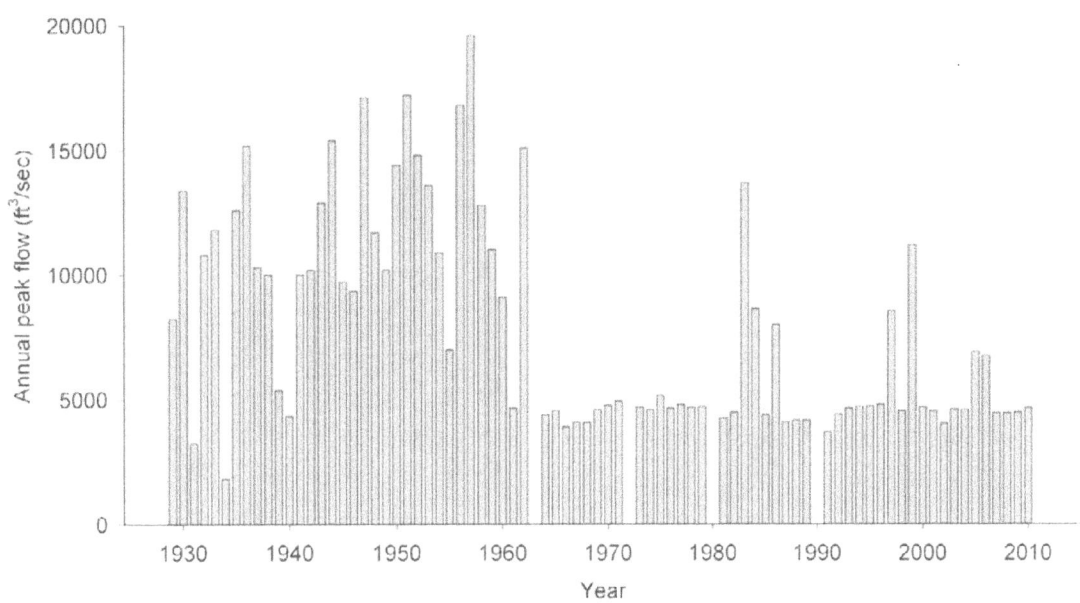

Figure 4-25. Instantaneous peak discharge of the Green River at the Greenwood and Lindale gauges, Utah.

USDA Forest Service Gen. Tech. Rep. RMRS-GTR-282. 2012

71

72

USDA Forest Service Gen. Tech. Rep. RMRS-GTR-282. 2012

Chapter 5: Linking Riparian and Wetland Vegetation to Hydrologic Factors

As discussed in Chapter 3, water is a principal limiting resource for plants, and a number of vegetation attributes are supported by and respond directly to water availability. In addition to direct measurements of plant characteristics at the level of the *individual* (such as water status, transpiration, water source, and incremental growth), attributes of vegetation may be measured at the *population* and *community* levels. Plant fitness, vulnerability to pathogens and herbivores, fecundity, competitive ability and productivity, population structure, and community composition and richness are influenced by water availability in space and time. Each vegetation attribute that could be measured at each of these levels varies in what it reveals about the characteristics of the vegetation. Each attribute also varies in its sensitivity to altered water availability and its interrelationship with other environmental and biological factors not associated with flow regime (Table 5-1). Most plants have optimum soil water conditions for growth that lie somewhere between saturated (and anoxic) and dry. This may change over the life span of the species (different for seedlings compared to adults) and over the season (e.g., higher during the height of the growing season compared to when plants are dormant). In this section, we discuss how to link the measurable characteristics of plants and vegetation (Chapter 3) to surface water, soil water, and groundwater measurements (Chapter 4). These tools are useful in modeling the distribution and fitness of plants and characteristics of vegetation along moisture gradients; making educated predictions of likely changes in response to altered hydrologic regimes; and informing decisions on proposed water extraction, groundwater pumping, and prescriptive and managed hydrologic regimes along rivers, streams, marshes and lakes.

Table 5-1. Metrics of riparian vegetation and sensitivity to hydrologic alteration and ability to reflect responses to chronic changes in flow regime. Number of asterisks indicates the authors' conceptions of relative strength. From Merritt and others (2010a).

Organizational level	Metrics	Acute sensitivity to hydrologic alteration	Reflective of chronic hydrologic alteration
Individual	xylem water potential	****	***
	transpiration	****	*
	photosynthesis	****	*
	CO_2 flux	****	*
	canopy volume	****	*
	shoot/root growth	***	*
	incremental growth	**	****
	leaf size	**	**
	leaf thickness	**	**
Population	age/stage/size class distribution	*	****
	population growth rate	*	****
	variability	*	****
Community	richness	*	**** (varies)
	diversity	*	**** (varies)
	composition	**	****
	cover	***	****
Stand Structure/ Productivity	biomass	***	****
	vegetation volume	**	****
	vertical structure	**	****

Vegetation response to changing hydrologic regimes may be directional such as the decline and collapse of existing plant populations, the establishment of new populations, or shifts in the zonation of vegetation. Response may also involve more complex and less directional shifts or abrupt changes in individual fitness, population structure, or species composition. A number of plant responses to altered water availability occur over different time scales ranging from hours to decades, and there may be lag times between hydrologic alteration and plant responses. Furthermore, different life stages of the same species may respond in different ways to the same hydrologic alteration.

In general, measures of the physiological characteristics of *individual plants* may be the most sensitive indicators of short-term changes in hydrologic regime, but may reveal little about the long-term ecological consequences of changes in water availability. Measures of the attributes of a *population* of a particular species or multiple species at the reach scale may better reveal changes in flow regime over longer periods of time, but may be less sensitive to subtle changes. In turn, attributes measured at the scale of the *plant community* may more reliably reveal long-term patterns of hydrologic regime. The choice of measurement at the scale of the individual, population, community, or some combination largely depends upon the vegetation attributes that are deemed important along with established goals for maintenance and restoration of wetland and riparian vegetation. Measures of plant attributes at each of these levels of organization as well as methods for understanding linkages between hydrology and individual, population, community, and functional attributes of vegetation are presented in the following sections.

Individual plants

Individual plants must germinate/sprout, grow, mature, reproduce, and die to contribute to a population (Figure 5-1). The likelihood of transitioning from one stage to the next in this cycle depends on the environment and the individual's traits for dealing with the environment. The same environmental factor may pose a different risk to individuals at each stage in development, influencing the rates of birth, growth, maturation, fertility,

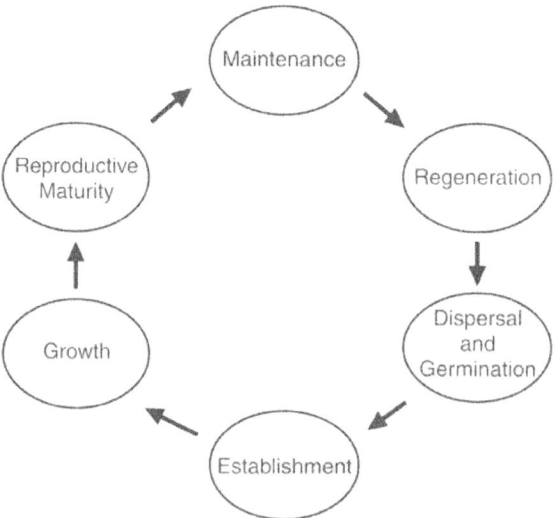

Figure 5-1. Life cycle of plants. In order to sustain a population, individual plants must successfully transition from one stage to the next over some time interval that is less than the maximum life expectancy of the plant.

and mortality. Knowledge of the individual life history (requirements and tolerances at different life stages) of a species relative to the physical environment is necessary for developing cause-and-effect relationships between attributes of hydrology and life stages of that species (Figure 5-2). Such information can be gleaned from peer-reviewed literature, textbooks, online databases, and/or personal experience and field measurements.

Any of the fitness variables measured on individual plants may be linked to hydrologic characteristics using standard statistical techniques. For example, xylem water potential or photosynthesis may be related to depth to water or streamflow volume using regression. Measurements of water stress of an individual (or several individuals) may be taken while water availability is changing (e.g., groundwater pumping causing depth to groundwater to change). For example, Cooper and others (2003b) measured xylem water potential of several cottonwood trees while groundwater pumping was causing lowering of water tables (Figures 5-3 and 5-4). Regression revealed relations that could then be used to provide guidance on how deep the groundwater could be pumped before dangerous or mortal stress levels were reached.

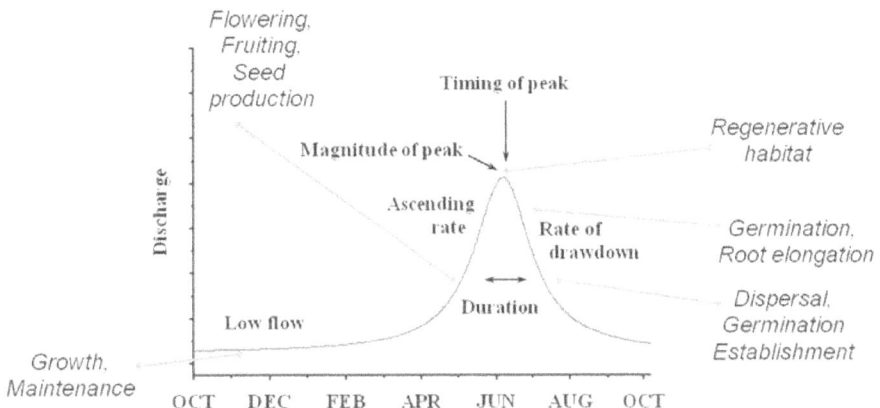

Figure 5-2. Linking each stage in the life cycle provides tools for managing water availability to provide for survival at each of the life stages and to accommodate transition from one stage to the next.

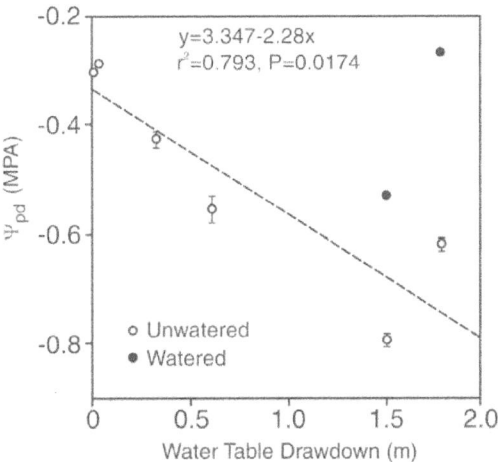

Figure 5-3. Linear regression of predawn xylem pressure potential along the maximum water table drawdown gradient for unwatered plots. Watered plot means are also shown. Error bars are ±1 SE. From Cooper and others (2003b).

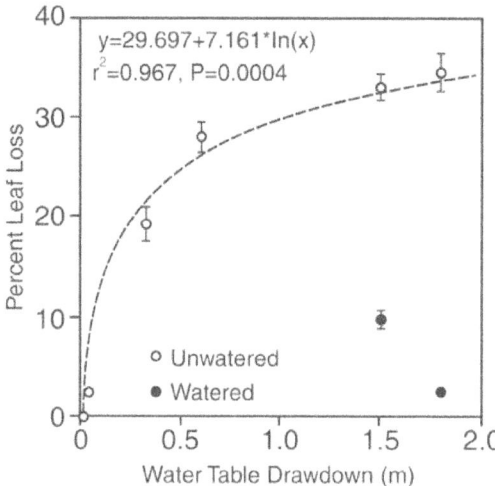

Figure 5-4. Polynomial regression of percent leaf loss along the maximum water table drawdown gradient for unwatered plots. Watered plot means are also shown. Error bars are ±1 SE. From Cooper and others (2003b).

Scott and others (1999) investigated the relations between tree survivorship, tree crown volume (measured as a percent change over time), incremental stem growth, and branch growth and depth to water table (Figure 5-5). They found that crown volume is tightly linked to depth to groundwater and that tree mortality is a function of the previous year's crown volume. They used logistic regression to model the probability of tree survival as a function of previous year's crown volume and found that trees with no change in crown volume had a 97 percent chance of survival, whereas those that experienced 30 percent or more loss in crown volume had less than a 50 percent chance of survival.

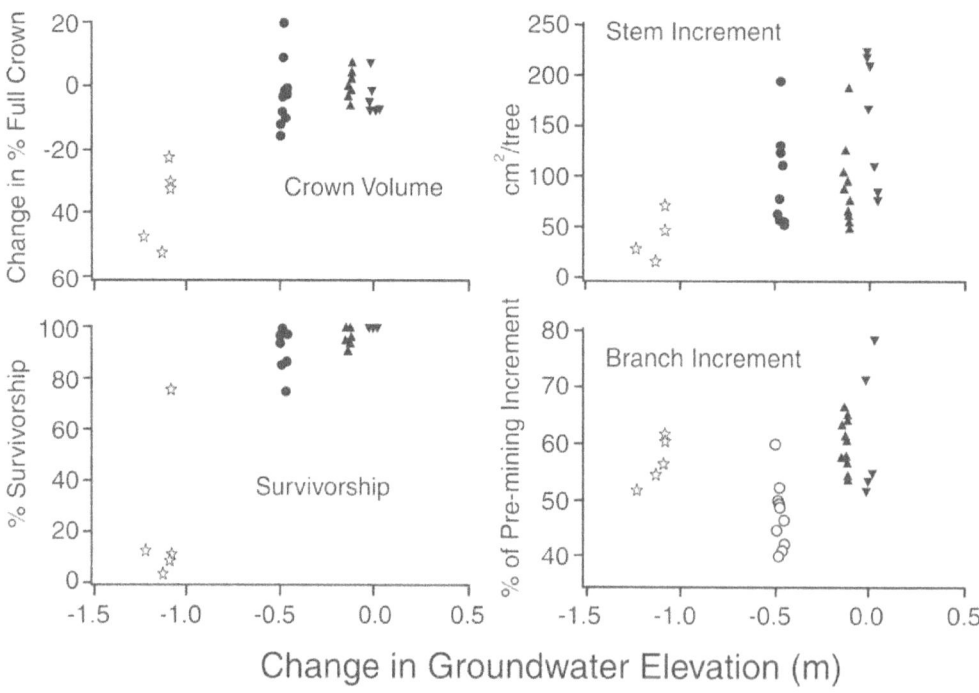

Figure 5-5. Measured attributes of cottonwood as a function of change in water availability (change in groundwater elevation) during groundwater pumping associated with mining. From Scott and others (1999).

USDA Forest Service Gen. Tech. Rep. RMRS-GTR-282. 2012

Linear regression can be performed in any standard statistical software, many plotting software packages, and spreadsheet programs such as Microsoft Excel. Logistic regression models binary (e.g., presence-absence or survival-mortality) data measured along a gradient. Generalized linear models (GLM) (e.g., logistic regression for presence-absence data and Poisson regression for count data) can be fitted in standard statistical software programs. Add-ons to Microsoft Excel may also be used to conduct generalized linear modeling (e.g., http://sunsite.univie.ac.at/Spreadsite/poptools/index.htm or http://www.statisticalengineering.com/glm.htm).

In addition to modeling plant attributes as a function of hydrologic variables, comparisons between control and treated areas may also be conducted. One such comparison could be upstream to downstream from a diversion or in an area of interest (treatment) versus an area used for comparison (control). In a study of a diverted creek in Arizona, xylem pressure was found to be lower (more water stress) in native trees along a reach that had been dewatered as compared to a free-flowing reach upstream from the water diversion (Figure 5-6).

If relationships between attributes and water availability are strong (e.g., high r-square and low p-value), they may be used to predict plant conditions under a variety of water availability scenarios. If the relationships are weak, it probably means that other factors (hydrologic and otherwise) are more important than the one measured. If multiple hydrologic factors are thought to be influencing plant characteristics, multiple regression analysis may be used to model conditions as a function of multiple variables.

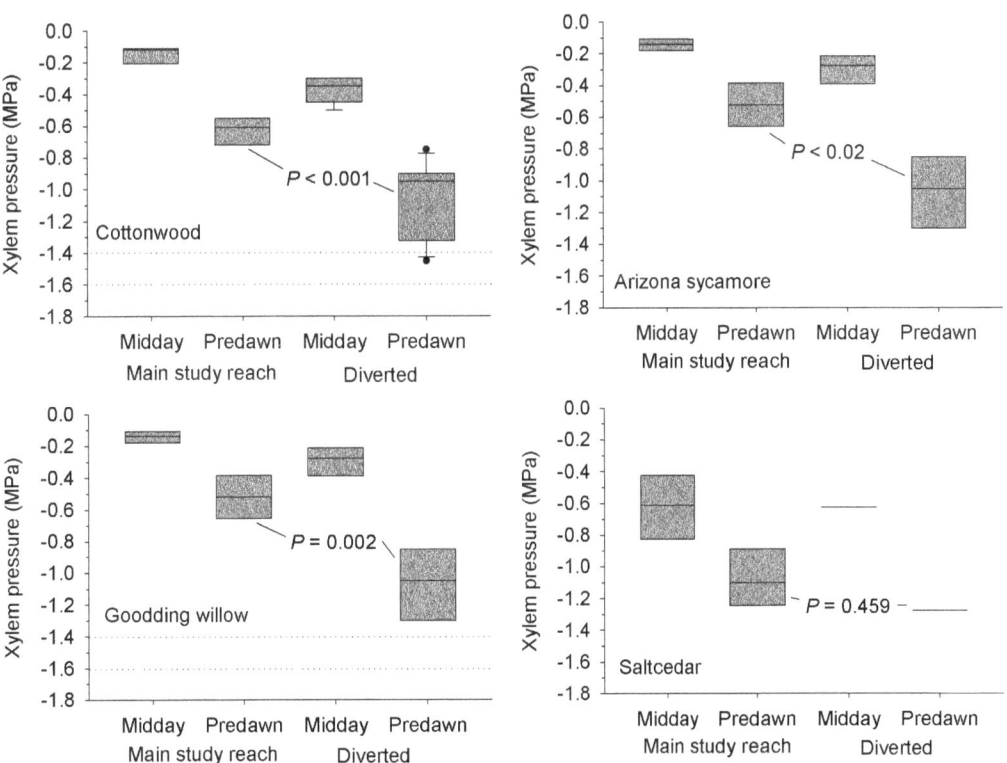

Figure 5-6. Box plots of pre-dawn and midday xylem water potential for three native tree species (*Platanus arizonica, Populus deltoids*, and *Salix gooddingii*) and a non-native (*Tamarix* spp.). T-tests comparing the mean midday water potential upstream and downstream from a diversion indicated that the native species all experienced low water potentials downstream from the diversion, but the non-native species exhibited drought tolerance.

Plant recruitment and establishment

Recruitment into a population first requires that seeds or vegetative propagules reach suitable sites, germinate, and survive. In many wetlands, well-developed seed banks can emerge when conditions permit and sites may be colonized by adjacent vegetation or propagules from elsewhere. Because the germinant to seedling stage may be the most vulnerable in the life history for most species, an understanding of each plant's requirements for germination and survival (termed the "regeneration niche" by Grubb [1977]) is particularly important in managing site hydrology to encourage recruitment of desirable species.

The requirements for establishment of many wetland and riparian species are understood, and quantifications of conditions that increase the probability of survival have been derived from tracking the fates of individuals in field studies (Houle 1994, Johnson 2000, Dixon 2003) and through experimentation (Horton and others 1960, Scott and others 1993, Segelquist and others 1993, Horton and Clark 2001).

Along rivers, flow-related factors such as fluvial disturbances create open patches for establishment of colonizing species, wetting of soils, and deposition of sediment. Fluvial processes may create and maintain open sites for recruitment through scour and deposition associated with the meandering processes, braiding, bar formation, overbank flooding, and lateral and vertical deposition of sediment (Scott and others 1996, Cooper and others 2003a). Whereas generalist recruitment strategies may enable some species to establish successfully over a range of conditions, many pioneer species in wetlands and riparian areas are reproductive specialists, requiring open substrate and specific moisture requirements for successful recruitment (e.g., the family Salicaceae) (Karrenberg and others 2002). The rate of surface water and groundwater level decline during initial establishment has been shown to be important for a range of species (Hughes and others 1997, Dixon 2003, Rood and others 2007).

Though there are few formal rules for linking the regeneration niche to flow regime along rivers, the wealth of information about cottonwood autecology, specifically recruitment, has led to the development of some very useful tools for managing river flows. One such tool is the "recruitment box" model (Mahoney and Rood 1998) (Figure 5-7).

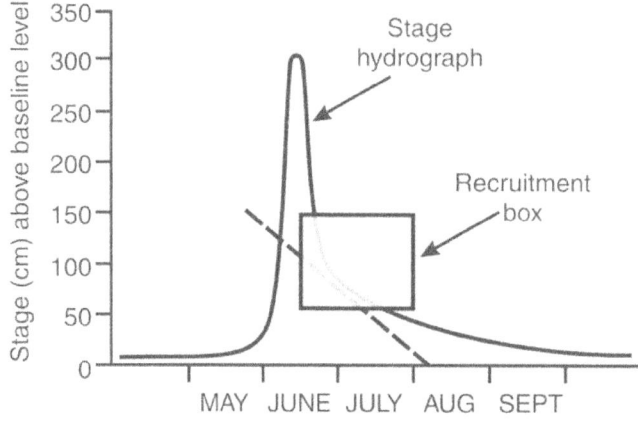

Figure 5-7. Recruitment box model developed by Mahoney and Rood (1998). The model incorporates the optimal timing and rate of drawdown of river stage-groundwater for cottonwood seedlings, but the concept is applicable to other species that are dependent on declining water table and that disperse their seeds in association with specific hydrograph attributes.

The model formalizes relationships between cottonwood root growth rate and surface and groundwater regime. The model integrates the timing of cottonwood seed release, the range of river stages that define the optimal position on the floodplain for seedling survival (high enough on the floodplain to avoid scour by subsequent floods; low enough to avoid drought stress), and the rate of stage/groundwater decline suited to maximum cottonwood seedling root extension rates (~2.5 cm/day). Along flow regulated rivers, the model can be used to manage hydrologic regimes (magnitude of peak, timing of peak, and rate of stage decline) to enhance cottonwood seedling survival.

The recruitment box model has been widely used to aid in the design of flow regimes to enhance recruitment for riparian forest restoration (Rood and others 2003, Rood and others 2005). The model assumes that river stage and alluvial groundwater decline are closely coupled, which may not be the case along gaining river reaches, in fine textured substrate, and in sites with complex substrate stratigraphy (Cooper and others 1999, Merigliano 2005). Seedlings in well-drained soils may be more vulnerable to rapid groundwater decline than those growing in finer textured substrate (Cooper and others 1999).

Although this elegant model was developed for cottonwood in western North America, the concept has also been applied to *Salix* spp. (riparian willow; Rood and others 2005) and could be readily transferable to other sexually reproducing species with specialized recruitment traits. The model could also be used to prevent invasion by undesirable species by decoupling the timing of seed release from the availability of suitable habitat (Shafroth and others 1998). This model is not applicable to clonal plants because they may reproduce largely asexually, for example *Populus angustifolia* (narrow leaf cottonwood) and *Salix exigua* (sandbar willow).

Of course, the survival of an individual or patch of individuals depends on other factors such as current and future conditions of the geomorphic features upon which they are deposited and germinate. Factors such as light availability, nutrient availability, substrate texture (and water holding capacity), presence of and competition with other individuals or species, herbivory, and flow related disturbance such as flooding and scour or burial are all important in determining the success of establishment (Francis and Gurnell 2006). However, the availability of water in appropriate amounts and at the correct times for individual species or groups of species increases the likelihood of their survival.

Growth and maintenance

Once individuals have successfully passed from seedling to juvenile stage, hydrologic processes (e.g., water availability) continue to be a determinant of growth, long-term survival, and mortality for wetland and riparian plants (Stromberg and Patten 1990, Cooper and others 2003a). As mentioned above, physiological and morphological attributes (e.g., water status, photosynthesis, and transpiration) of riparian plants tend to be the most sensitive to changes in flow regime over short time scales (Table 5-1). Physiological responses to changing hydrologic conditions can occur over the period of several hours and can have long-term repercussions for the morphology and fitness of the individual. Water stress and reduced transpiration can result in lower growth rates, which in herbaceous species can be measured during the growing season, and in woody vegetation can be measured over multiple seasons.

Measurements of incremental growth such as annual branch growth and tree ring width has been very effectively modeled as a function of total annual flow volume or specific streamflow attributes along streams (Stromberg and Patten 1996, Disalvo and Hart 2002) (Figure 5-8). Such relations can be used to set targets for total annual flow volume to promote some desirable or acceptable growth rate of trees and to avoid volumes that inhibit growth.

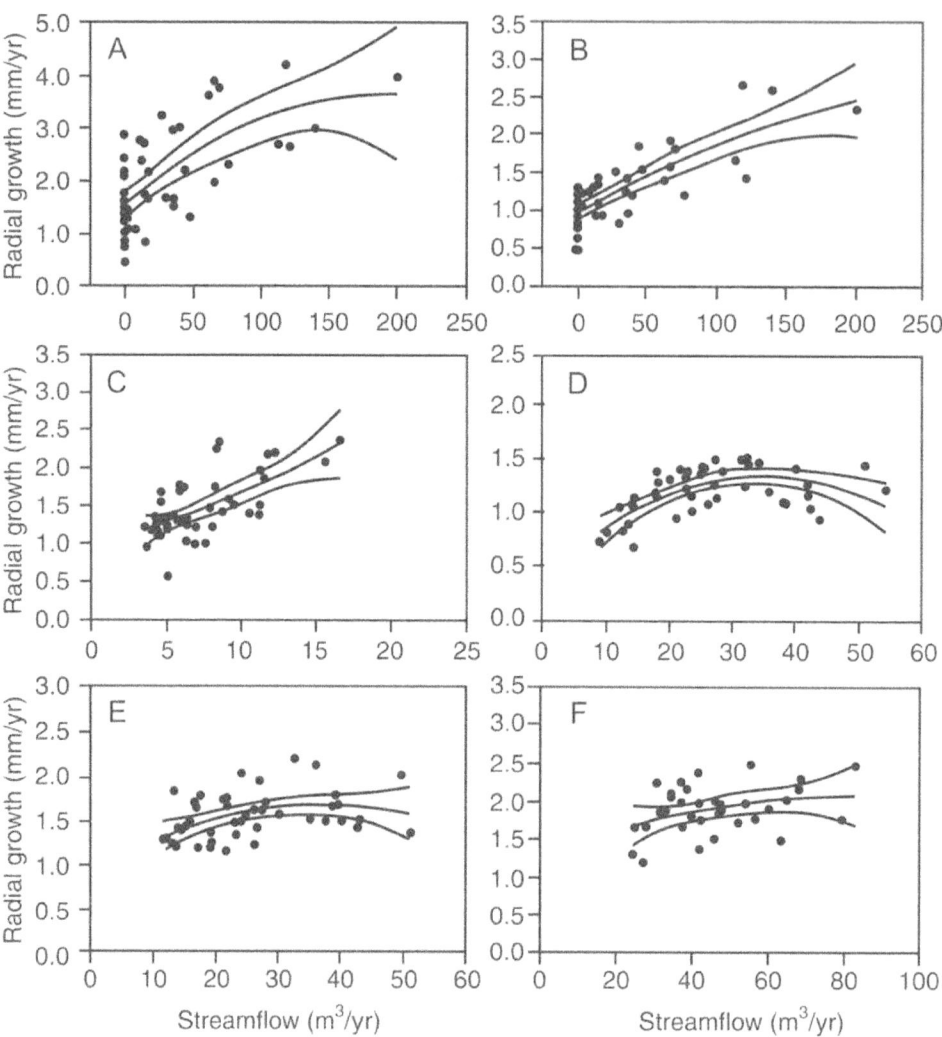

Figure 5-8. Radial growth of cottonwood (*Populus trichocarpa*) as a function of streamflow for three stream types: (A, B) diverted streams in wide, alluvial valleys; (C, D) free-flowing streams in wide, alluvial valleys; and (E, F) free-flowing streams in narrow, confined valleys. Regression lines and 95 percent confidence intervals are indicated. From Stromberg and Patten (1996).

Individual distributions

Flow-related variables have universally been found to be strong predictors of species distributions in wetlands and riparian ecosystems (Franz and Bazzaz 1977, Shipley and others 1991, Merritt and Cooper 2000). Species are sorted along elevation and microtopographic gradients according to differences in their response to flooding regime (Franz and Bazzaz 1977, Auble and others 1994). When related to environmental conditions (e.g., soil moisture, soil texture, nutrients, depth to water table), the present distributions of individual species on a landscape can aid in understanding the range of conditions under which they can survive. It is also useful to examine where they do not occur as that may indicate that the plants are stressed or outside of their tolerance for some resource or stressor. The niche of a species may be represented as a multidimensional representation of biotic and abiotic factors to which the population responds. Fitting a distribution or function to presence-absence or abundance data of the species along an environmental gradient (e.g., flow related) or multiple gradients is a way to simultaneously test a hypothesis about its niche and develop a framework for prediction of its distribution under changing conditions (Franz and Bazzaz 1977).

Distributions or functions are fitted to plot-level or site-level presence-absence or abundance data most commonly without regard for stage of development in plants ("holistic" approach), as opposed to examining the characteristics of different age or size classes. If plants are distributed linearly along a gradient (usually because only a portion of the gradient was sampled), linear or log linear regression can be used to fit a curve to the abundance data. If presence-absence data are being analyzed, probit or logit (logistic) regression is used to model probability of presence. If a wide gradient is sampled and species are unimodally (bell shaped or normally) distributed, Gaussian regression is usually used to model abundance data, and polynomial logistic regression is used to model probability of occurring along the gradient(s). The steps in the process of fitting Gaussian (normal) curves to species abundance data involves:

(1) Plotting species abundance data along a hydrologic gradient comprised of hydrologic values for each corresponding vegetation sample. Hydrologic variables may include soil moisture, flow duration, flood frequency, depth to groundwater, or any number of factors that might explain variability in distributions of individual plant species;

(2) Fitting an appropriate curve (model) to the data (Figure 5-9); and

(3) evaluating the fit of the curve to the data.

Such species distributions provide insight into the breadth of the species' realized niche (Hutchinson 1957), indicate its "ecological amplitude" (affinity and tolerance to resources and stressors), and its environmental optima along a hydrologic or other gradient(s).

Curves may be fitted in any plotting software or statistical program (e.g., SAS, SigmaPlot, R) or with curve fitting add-ons for Microsoft Excel. If model fit statistics (r-square, log likelihood, Chi square goodness of fit, etc.) indicate that the model adequately describes the data, these models may be used to predict abundance in unmeasured sites or under modified conditions. General methods for fitting models are reviewed by Jongman and others (1995).

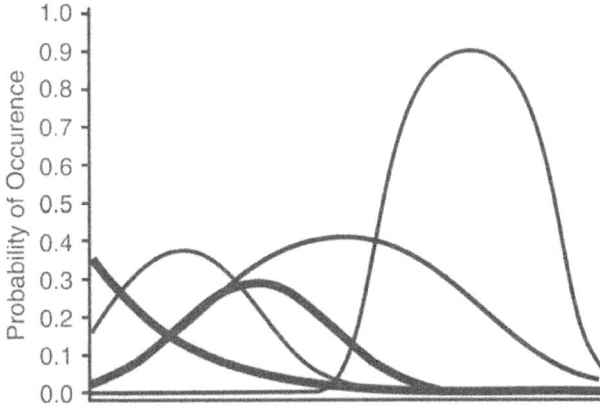

Figure 5-9. Probabilistic species response curves. The independent variable may be any of a number of flow-related gradients such as soil moisture, depth to groundwater, inundation duration, flood frequency, or any number of plant-relevant hydrologic variables. Different lines represent different species.

Community Analysis

An alternative to probabilistic modeling of each species is indirect and direct gradient analysis using a variety of ordination techniques (Jongman and others 1995). Ordination can reduce the dimensionality of the data so that plots are arrayed in a way that may be plotted in two or three dimensions. Vegetation plots are arrayed in the ordination space so that compositionally similar plots are closer together and those that are less compositionally similar are further apart. Principal components analysis (PCA), detrended correspondence analysis (DCA), non-metric multidimensional scaling (NMDS), redundancy analysis (RA), and canonical correspondence analysis (CCA) are all options for reducing multivariate species composition data from many plots into fewer dimensions for analysis and visualization. DCA and CCA assume unimodal (Gaussian) distributions of species and array these distributions along axes either defined by combinations of the species (DCA) or constrain the ordination axes to be linear combinations of measured environmental variables (CCA). PCA may be used when only a portion of a gradient is sampled and the species are distributed linearly along the portion sampled (Figure 5-10). When assumptions regarding multivariate normality are violated, NMDS may be used as a non-parametric alternative.

Regressing flow-related variables against PCA or DCA axis scores can reveal which variables the species are collectively organized along and may provide an indication of which flow-related variables best account for variation in community structure (Vanderijt and others 1996). By constraining the ordination axis scores to be linear combinations of measured environmental variables, CCA also provides tools for identifying the principal flow-related variables describing species distributions (Figure 5-11). An assumption of CCA is that all important variables have been included in the model. These techniques, which have become widely used in community ecology, bridge the gap between analyzing individualistic responses of species and examining patterns in community structure.

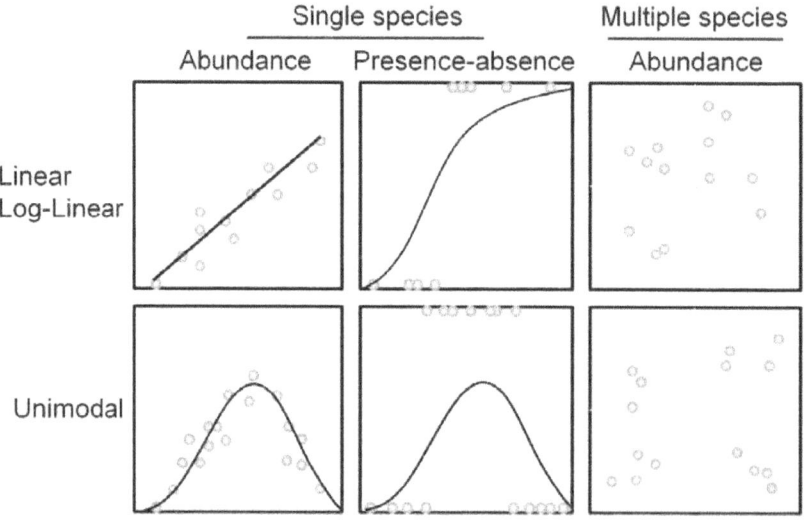

Figure 5-10. Models fitted to species abundance and presence-absence data along a hydrologic gradient (single species models). Multiple species plots are from ordinations and the first two ordination axes are shown. From upper left in a clockwise direction: linear regression, logistic regression, principal components analysis, detrended correspondence analysis, polynomial logistic regression, and Gaussian function.

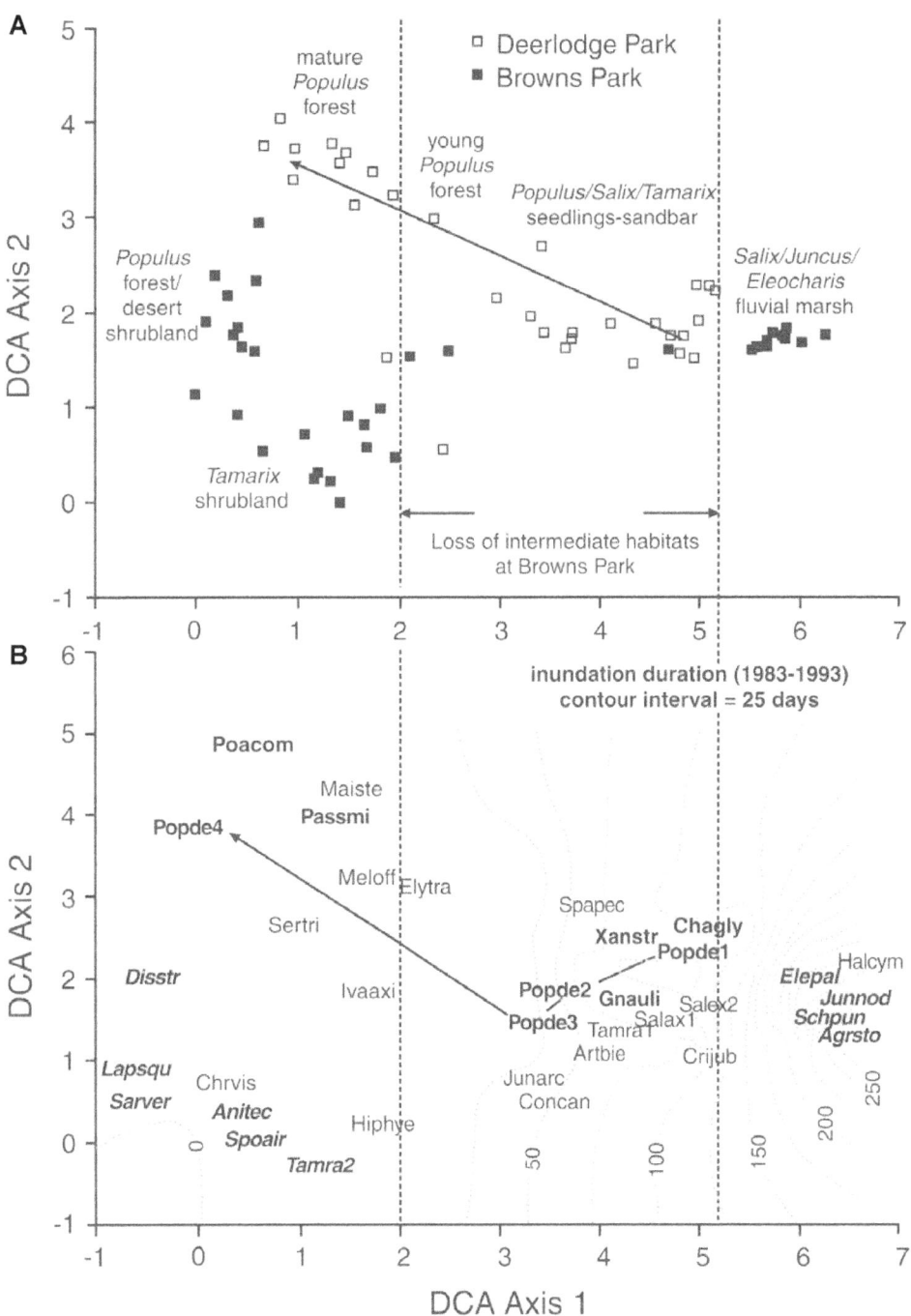

Figure 5-11. Plots of the first two axes from detrended correspondence analysis of vegetation samples taken in riparian areas along the free-flowing Yampa River and the regulated Green River in Colorado and Utah, respectively. The top plot (A) illustrates a successional sere dominated by cottonwood forest with gradual species turnover along the free-flowing river and an abrupt transition between dry uplands and fluvial marsh along the regulated river. The lower plot (B) illustrates the distribution of several species (indicated by six letter acronym) arrayed along an inundation duration gradient. From Merritt and Cooper (2000).

PCA and NMDS may be performed in many statistical programs, however DCA and CCA are specialized and are only available in community analysis software programs such as PC-Ord (http://home.centurytel.net/~mjm/), CANOCO (http://www.pri.wur.nl/uk/products/canoco/), and Multi-Variate Statistical Package (http://kovcomp.com/mvsp/index.html). The vegan package of the statistical computing software package R can perform many of the analyses outlined above, including DCA, CCA, PCA, and NMDS (available for free download at http://cran.r-project.org/web/packages/vegan/index.html).

The vegetation data necessary to perform analysis by any of these approaches is typically in the form of plot numbers (unique identifier) in the first row and species names (typically acronyms) as row headers with individual cells in the matrix containing an abundance value (e.g., percent cover) for each species. If the species is absent from a particular plot, a zero value is entered (Figure 5-12).

A wide range and combination of flow-related variables may be used in gradient analysis to define the niche space of each species at a site or to include in ordination. Such applications of direct gradient analysis assume that the relationships between the environmental gradient and the distributions of plant species are inherent attributes of the species and do change as a result of altered hydrology. Quantitatively relating flow

	A	B	C	D	E	F	G	H	I	J	K	L	M	N	O	P	Q	R	S	T	U	V	W
1	Species	111	112	121	122	123	131	132	141	143	144	145	211	213	214	215	216	217	218	219	2110	221	223
2	*Abies concolor*	0	0	0	0	0	0	0	0	0	0	5	0	0	0	0	0	0	0	0	0	0	0
3	*Acer negundo*	0	0	0	0	0	0	0	0	0	0	20	0	0	0	0	0	0	0	0	0	0	0
4	*Achillea millefolium var. lanulosum*	0	2	2	7	3	0	1	1	0	0	7	0	2	5	8	1	0	1	0	1	3	2
5	*Agastache pallidiflora ssp. neomexicana*	0	0	0	0	0	0	0	0	0	0	0	0	0	0	0	0	0	0	0	0	0	0
6	*Agrimonia striata*	0	0	0	0	0	0	0	0	1	0	0	0	0	0	0	0	0	0	0	0	0	0
7	*Agrostis stolonifera*	35	15	10	15	15	0	0	0	0	0	0	0	0	0	0	0	0	0	0	0	0	0
8	*Allium cernuum*	0	0	0	0	0	0	0	0	0	0	0	0	0	0	0	0	0	0	0	0	1	0
9	*Anaphalis margaritacea*	0	0	0	0	0	3	0	1	0	1	0	0	0	1	0	1	0	0	0	0	1	1
10	*Artemisia dracunculus*	0	0	0	0	0	0	2	0	0	0	2	0	2	0	0	0	0	0	0	0	0	0
11	*Artemisia ludoviciana*	0	20	25	1	3	0	0	0	3	20	5	1	2	2	7	0	0	4	0	0	0	5
12	*Bahia dissecta*	0	0	0	0	0	0	0	0	0	0	0	0	0	0	0	0	1	1	0	0	0	0
13	*Bromus anamalous*	0	0	0	0	0	9	0	0	0	0	0	0	0	0	0	0	0	0	0	0	0	0
14	*Bromus carinatus*	0	0	0	20	0	0	0	0	0	0	0	0	0	0	0	0	0	0	0	0	0	0
15	*Bromus ciliatus*	0	0	4	5	22	0	0	1	0	4	7	0	3	15	3	3	0	0	0	0	4	0
16	*Bromus inermis*	0	3	0	0	0	0	0	0	1	0	0	0	0	0	0	0	0	0	0	0	0	0
17	*Carduus nutans*	5	3	0	0	0	0	0	0	0	0	0	0	0	0	0	0	0	0	0	0	0	0
18	*Chenopodium graveolens*	0	0	0	0	0	0	0	0	0	0	0	0	0	0	0	0	1	1	0	0	0	0
19	*Cirsium parryi*	3	0	0	0	0	0	0	0	0	0	0	0	0	0	0	0	0	0	0	0	0	0
20	*Commelina erecta*	0	0	0	0	0	0	0	0	0	0	0	0	0	0	0	0	0	0	0	0	0	2
21	*Conyza canadensis*	0	0	0	0	0	0	0	0	0	0	0	0	0	0	0	0	0	0	0	0	0	0
22	*Cosmos parviflorus*	1	0	0	1	2	5	2	3	0	0	0	0	0	0	1	1	0	0	0	1	0	2
23	*Cyperus spaerolepis*	0	0	3	1	0	0	0	0	0	0	0	0	0	0	0	0	1	0	0	1	0	1
24	*Dactylis glomerata*	0	0	15	20	15	4	1	0	0	2	10	0	1	5	0	2	0	0	0	0	8	0
25	*Dalea polygonoides polygonoides*	0	0	0	0	0	0	0	0	0	0	0	0	0	0	0	0	0	0	0	0	0	0
26	*Elymus trachycaulis*	0	0	3	10	0	0	0	0	0	0	0	0	0	0	0	0	0	0	0	0	2	0
27	*Erigeron subtrinervis*	0	0	0	0	0	1	0	0	0	0	0	0	0	1	0	0	0	0	0	0	8	1
28	*Euphorbia albomarginata*	0	0	0	0	0	0	0	0	0	0	0	0	0	0	0	0	1	0	0	0	0	0

Figure 5-12. Plot by plant species matrix in Microsoft Excel for input to many statistical analysis software programs.

attributes to response of individual species provides a continuum of probable change. Collectively, the flora in a riparian area may contain a range of species with numerous combinations of tolerances and requirements, some of which are in direct conflict. Niche-based modeling quantifies this and provides an estimate of how different flow regimes in different years could provide for the requirements of a range of species. Another important consideration of niche-based modeling is that the response of different species to changes in flow regime varies widely. Short-lived species such as annuals could respond on the scale of a single season, but longer-lived species could require decades or even centuries to completely respond. Furthermore, the predictions from niche-based models only provide an estimate of the probability of change. A variety of other factors (fire, herbivory, competition, disease, fluvial disturbance) influence the distributions of species, and if not explicitly incorporated into models, these factors increase the uncertainty of predictions (Auble and others 2005). These models provide a probabilistic view of responses. These probabilities of change may be used as a risk assessment to inform the development of limits on permissible hydrologic alteration.

Managing for single species is rarely advisable; however, individually modeling a large number of species can be problematic. Direct gradient analysis has been applied not only to probabilistic modeling of individual species and groups of individual species but also to modeling of plant associations (communities) and cover types, as will be discussed later in this chapter.

Populations

Often, river management plans are tailored to keystone, indicator, or umbrella species, or species of special concern (e.g., threatened or endangered) (Lambeck 1997). It is important not only to know the distribution of the species across the landscape, but also to understand how the species is performing (e.g., recruitment and overall fitness) at the population level. This can be accomplished through characterizing the population structure using age-class distributions or modeling populations using structured modeling.

Examining structure (age- or size-class distributions) of a population is helpful in determining what life-stage is most affected by changes in flow regime. Age- or size-class distributions, evidence of successful recruitment, or some functional assessment of life-stage can yield significant insight into the structure of the population and can provide an indicator of "bottlenecks" that may be negatively affecting one or more life-stages (e.g., recruitment failure). Age-class distributions may also provide insight into event-driven recruitment events that may be statistically related to hydrologic characteristics (Figures 5-13A and B) (Auble and Scott 1998, Birken and Cooper 2006, Rathburn and others 2009). Such distributions can be statistically compared between two different populations or for a single population to document the population-level response to changing environmental conditions.

As mentioned, tolerance of plants to inundation, fluvial processes (scour and burial), anoxia, and drought vary as a function of developmental stage for many species adapted to riverine environments (Smith and others 1998, Friedman and Auble 1999). In riparian areas, many colonizing species germinate and establish on freshly deposited alluvium near the channel, whereas adults persist as channel migration and floodplain abandonment cause local soils to become more desiccated over time. Thus, a species may be a specialist requiring specific conditions to establish and exhibit more generalist traits during later stages of development. The better we understand the specific relationships between flow-related processes and the survival and mortality of plants at different stages, the greater an opportunity we have for modeling populations and examining differences in population growth rates in response to different combinations of environmental factors.

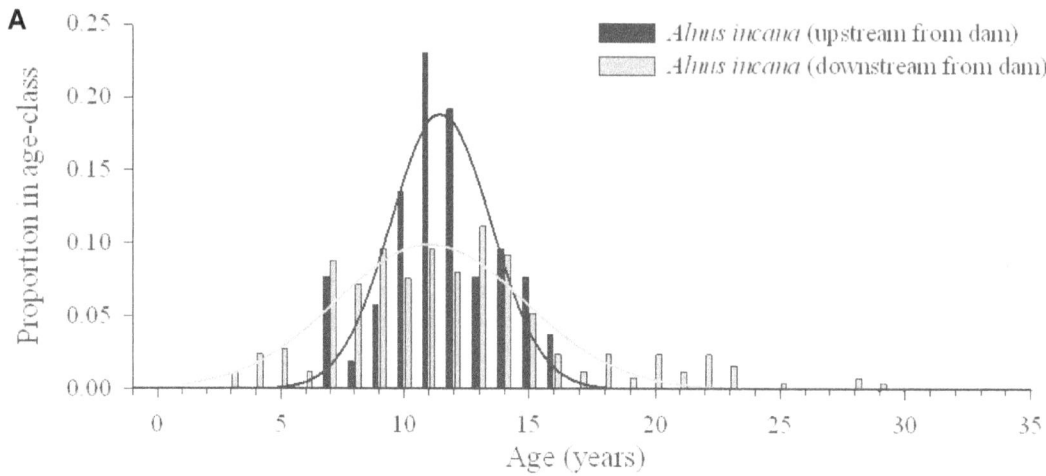

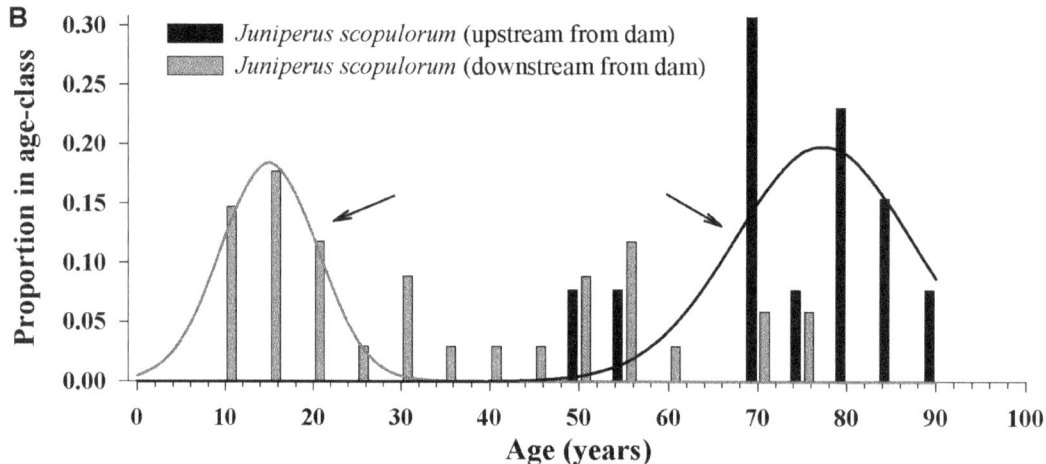

Figure 5-13. Age class distribution for alder (*Alnus incana*, A) and western red cedar (*Juniperus scopulorum*, B) upstream and downstream from a dam along the North Fork Cache la Poudre River in Colorado (Rathburn and others 2009). These distributions indicate that size class distributions of alder are similar upstream and downstream from the dam. In contrast, there is a wide distribution of ages of western red cedar (an upland species) in riparian areas downstream from a dam compared to the narrow distribution of older individuals upstream from the dam. The divergence of upstream and downstream population structures is accentuated by the arrows. This is one indication that regulated flows may be enabling encroachment of western red cedar into riparian areas downstream from the dam.

Though formal approaches for modeling and projecting population dynamics of species through time may be very instructive, the techniques are mathematically intensive and are beyond the scope of this guide (Caswell 2001, Lytle and Merritt 2004).

Plant Communities

Hydrologic alteration often results in shifts in predictable community-level attributes of riparian vegetation, including species richness and plant community composition (Nilsson and others 1991, Jansson and others 2000, Merritt and Cooper 2000). Collective attributes of the community (e.g., richness, diversity, cover and biomass) are linked to hydrologic attributes of rivers and may respond in predictable ways to specific hydrologic

USDA Forest Service Gen. Tech. Rep. RMRS-GTR-282. 2012

alterations (Figure 5-11) (Nilsson and Svedmark 2002, Lite and others 2005, Birken and Cooper 2006). The ordination techniques that were previously described may be very useful in examining shifts in communities through time, comparing unimpacted to impacted conditions, and providing insight into relationships between hydrological variables and community attributes (Figure 5-11). In addition, there are a variety of other techniques for evaluating community attributes: examining differences in species richness, categorizing species according to some shared trait or common functional attribute, comparing relative proportions of species to a reference stream or reach, and examining similarity and dissimilarity in species composition between populations of interest. Many of these may be used in conjunction with ordination, to evaluate specific questions, and to provide complementary information.

Species richness (number of species) may be compared by evaluating average plot-level richness between populations of interest using t-tests to compare two groups or analysis of variance ANOVA for comparing more than two groups. When statistical assumptions for t-tests and ANOVA are not met (e.g., data are normally distributed), non-parametric tests such as Mann-Whitney U test and Kruskal-Wallis ANOVA, respectively, may be preferable.

Because number of species increases as a function of area sampled, care must be taken that the number of plots and their dimensions are the same between the populations being compared or that adjustments are made to account for differences in the area sampled. One such adjustment is to divide species richness in each plot or sample unit by the natural logarithm of the collective area of plots and to then compare area-adjusted richness, as previously described. Another method is to estimate the richness of an area by fitting species accumulation curves to the plot-level data from the area. Species area curves may be fitted using bootstrapping techniques in software such as Primer-E (http://www.primer-e.com/) or Estimate-S (http://viceroy.eeb.uconn.edu/estimates). Total species richness between sites may then be compared using Chi-square tests. Another simple method for comparing sites with different numbers of plots is to randomly select a subset of plots from the site that has the most plots so that identical numbers of plots are being compared in an unbiased way.

In addition to comparing richness, community biodiversity may be compared using one of many biodiversity indices, which account not only for the number of species but the evenness of species as well. If average plot-level diversity is being compared, different numbers of plots may be used, but they must be of the same size. If total biodiversity is being compared, similar area (cumulative area of plots) must be sampled, as when comparing richness.

Another commonly used approach to compare community attributes is to classify species into functional groups based upon some shared trait or function (see Merritt and others 2010). The relative proportions of native to non-native species may also be compared. The life form of species may be grouped and compared between sites, treatments, or over time. For example, the relative proportions of trees; shrubs; herbaceous dicots; and grasses, sedges, and rushes may be compared, as these life forms and groups may respond in different ways to hydrologic alteration. Other functional grouping variables that could be used for community comparison are wetland indicator status (Reed 1988), Raunkiær life form (plants classified by where the perennating buds are stored on the plant) (Crawley 1986), Grime's (1977) plant strategy categories (competitor, stress tolerator, and disturbance-adapted), or any number of other relevant functional characteristics. For example, reduction in the relative proportion of obligate wetland species relative to upland species may provide a compelling quantitative measure of the effects of groundwater pumping or wetland draining on wetland plant communities. Lite and others (2005) classified plants into categories associated with their affinities to various depths to groundwater and reproductive traits (e.g., hydromesic perennial, hydromesic annual, xeric perennial, and xeric annual). Using the Raunkiær system of categorizing plants according to the position of buds or regenerating parts of a plant (Raunkiær 1934), Nilsson and others (1994) found differences in the traits of riparian plant communities between different sites. Communities may be compared for absolute or relative proportions of various functional categories using Chi-square tests.

Comparisons in species composition between two different communities or the same community before and after a treatment may be made by calculating a matrix of indices of community similarity or dissimilarity and comparing average similarity or dissimilarity within and between populations (Legendre and Legendre 1998). Such matrices may be constructed quickly from a plot-species matrix (Figure 5-12) using statistical programs (e.g., SAS) or specialized community analysis software such as Primer-E, MVSP, or R. Community similarity can be compared within and between populations using standard measures (e.g., Jaccard's measure, which uses only presence-absence; or the Bray-Curtis similarity coefficient, which accounts for abundance or others) (Legendre and Legendre 1998). Species accounting for differences may then be identified. Statistical comparisons may be made using t-tests or ANOVA to compare average similarity within and between populations. A more formal comparison may be made through performing an analysis of similarities (ANOSIM) (Legendre and Legendre 1998). ANOSIM has been widely used for testing hypotheses about spatial differences and temporal changes in assemblages and for detecting environmental impacts (Renöfält and others 2007). In comparing the degree of change from natural conditions anticipated for plant communities under various reservoir operation scenarios, Franz and Bazzaz (1977) predicted that vegetation dissimilarity would range from 16 to 30 percent for three different plans, and the authors recommended a flow regime that minimized departure from natural conditions. Specialized statistical software is available for performing ANOSIM using similarity indices chosen by the user (e.g., Primer-E).

Plant communities may also be compared after classifying them into cover types using either subjective classification or more objective classification. Plant communities may be objectively classified using divisive or agglomerative clustering using statistical software or specialized community analysis software such as PC-Ord, Two-Way Indicator Species Analysis (TWINSPAN; http://www.canodraw.com/wintwins.htm), Primer-E, MVSP, or R. In riparian areas and wetlands, plant associations (dominant cover types) show strong affinities for specific hydrologic attributes such as inundation duration (Franz and Bazzaz 1977, Auble and others 1994, Friedman and others 2006) and depth to groundwater (Rains and others 2004, Camporeale and Ridolfi 2006). In addition to examining individualistic responses of species to inundation duration, Franz and Bazzaz (1977) examined community response by analyzing probable changes in vegetation cover types as a whole. They used individual species response curves to predict the probability of occurrence of species in communities under various flow scenarios and then used dissimilarity indices to examine percent compositional departure from natural (reference) communities. Discriminant function analysis or Bayesian discriminant analysis may be used to classify plant cover types into independently determined cover types as functions of multiple environmental variables (Szaro 1990, Castelli and others 2000). These functions may then be used to predict shifts in cover type as a function of changing abiotic variables (Rains and others 2004).

Other methods of characterizing vegetation based upon overall stand attributes are: examining biomass, vegetation volume, and stand physical structure (Stromberg and Patten 1990, 1991; Stromberg and others 1993). These attributes can be regressed against hydrologic variables and the relationships used to estimate stand attributes in response to flow alteration, extraction scenarios, or climate change. One advantage of these simple approaches is that a full range of responses are modeled, enabling an evaluation of trade-offs between flow alteration and measurable riparian or wetland conditions.

Conclusions

A variety of methods are available for evaluating vegetation and its relationship to hydrologic conditions. Rather than a comprehensive review of the many methods that have been developed by plant ecologists over the past century, an overview of what we perceive to be some of the simplest and most effective approaches has been presented. Most of these approaches do not require a tremendous mathematical background or specialized skills to use, yet the approaches enable users to utilize the vegetation and hydrologic data gathered using the techniques outlined in other chapters of this guide to address specific questions, test specific hypotheses, make formal comparisons between sites or treatments, track trends over time, and develop predictive ability to inform management decisions.

The choice of one or a combination of organizational levels of plants (individual, population, community, or functional grouping) will hinge upon the management questions at hand. In some cases, a combination of sensitive and robust measures will be necessary (Table 5-1) to fully evaluate the causes of change or to project change under various flow alteration or water extraction scenarios. Tailoring the data collection and analyses to conform to management questions, legal mandates (e.g., threatened or endangered species and guidelines in land management plans), or a particular ecosystem type or set of stressors requires a basic understanding of the range of approaches presented here and some clear management objectives for species or communities of interest. One instructive way to formulate which approaches to apply is to examine examples from the literature and look for similarities to issues at hand. Table 5-2 provides a review of the approaches presented here on systems ranging from fens and small streams to large rivers, reservoir margins, and bottomland swamps. This review, examination of relevant references, and a basic foundation in the statistical techniques suggested here will provide tremendous opportunities for quantitative evaluation of plant-hydrology relations for management of wetland and riparian ecosystems of management concern. Some excellent references for additional explanation of the techniques presented in this chapter are: Manly 1994, Jongman and others 1995, Legendre and Legendre 1998, McCune and Grace 2002.

Table 5-2. Examples of studies examining riparian vegetation at individual, population, community, and guild levels. Hydrologic variables used to model vegetation attributes, statistical method, and assumptions of the various studies are given. From Merritt and others (2010a).

Model level	Citations	Hydrologic variable/s tested	Best hydrologic variables	Vegetation attribute	Analysis tool	Assumptions	Locations
Individual/ Establishment	Rood and Mahoney 1993; Mahoney and Rood 1998	Timing of peak, rate of stage decline	All	Seedling survival	Recruitment box model	Static channel geometry, coarse textured substrate	Oldman River, Alberta, Canada
Individual/ Maintenance	Busch and Smith 1995; Cooper and others 1999; Scott and others 1999	Depth to groundwater	Depth to groundwater	Xylem pressure potential, leaf thickness, leaf area, canopy volume, annual branch growth	Logistic regression, linear regression	Static channel geometry	Bill Williams and Coloraro R.; AZ, USA; eastern CO, USA; Green R., UT, USA
Individual/Cover types	King and others 1998	Water level	Water level	Stress, mortality, and regeneration, tree relative importance values	Probit analysis		Ouachita and Saline R., AR, USA
Population	Lytle and Merritt 2004; Grifith and Forseth 2005; Smith and others 2005	Distributions of floods, droughts, timing of peak, rates of change in flow; timing of flood, timing of drawdown	Distributions of floods, droughts, timing of peak, rates of change in flow; timing of flood, timing of drawdown	Population growth rate, stage-based population growth rate, sensitivities, elasticities, variability in population growth rate, aerial cover of lfe-stages	Stochastic structured/ matrix modeling	Quasi-equilibrium channel	Yampa R., CO, USA; Illinois R., IL, USA
Population	Clipperton and others 2003	Flow exceedance probability	n/a	Populus recruitment, growth, health (qualitative)	Qualitative coupling of requirements for recruitment, growth, and maintenance		Saskatchewan R. basin, Alberta, Canada
Population	Pearlstine and others 1985; Phipps 1979	Depth to ground-water, flood fre-quency, inundation duration	Depth to groundwater, flood frequency, inundation duration	Growth, dispersal, death, of five tree species	Numerical modeling	Static channel geometry	Santee R., SC, USA; White R., AR, USA
Community/ Individual	Franz and Bazzaz 1977; Auble and others 1994; 1998; 2005; Friedman and others 2006	Flow duration	Flow duration	Response curves of plant associations, response curves of individual species	Numerical modeling, response curves (Gaussian, logistic regression)	Static channel geometry	East-central IL; Gunnison R, CO, USA; San Miguel R, CO, USA
Individual/Stand	Stromberg and Patten 1990; Stromberg and Patten 1991	Annual flow volume, flow volume -1 and -2 yrs, cumulative flow pervious 4 years, season flow volume (Oct-Mar, Apr-Jun, Jul-Sept)	Annual flow volume the year of growth	Tree incremental growth, canopy vigor, mortality	Linear regression		Rush Creek and Bishop Ck., CA, USA
Individual/ Community	Stromberg 1993	Mean growing season flow volume, mean and median annual flow volume, flood magnitude	Growing season flow volume, flood magnitude	Abundance (foliage area, stem basal area, stand width), species richness	Second order linear regression	Static channel form	Verde R., Arizona, USA
Cover types/ Functional groupings	Rains and others 2004	Depth to groundwater, flooding	Depth to groundwater	Response curves of plant associations	Numerical groundwater model (MODFLOW), Bayesian classification	Static channel form	Little Stony Ck., CA

Table 5-2. Continued.

Model level	Citations	Hydrologic variable/s tested	Best hydrologic variables	Vegetation attribute	Analysis tool	Assumptions	Locations
Cover types	Johnson 1992	Water development		Change in riparian cover types	Compartmental simulation model/numerical modeling		Missour R., MI, USA
Cover types	Springer and others 1999; Baird and others 2005	Depth to groundwater		Woody vegetation cover, cover seedling establishment, cover juvenile survival habitat	Numerical groundwater model (MODFLOW) and conceptual vegetation model	Static channel geometry	Verde R., AZ, USA
Cover types	Primack 2000	Inundation duration (classes)	Inundation duration (classes)	Cover types	Cover types	Static channel geometry	Pere Marquette watershed, Michigan, USA
Community/ biomass	Camporeale and Ridolfi 2006	Stream discharge	Stream discharge	Probability of vegetation biomass	Stochastic modeling		Hypothetical
Stand character-istics	Perucca and others 2006	Distance from river (parabolic function to represent a position between anoxic and dry)	Distance from river (parabolic function to represent a position between anoxic and dry)	Biomass	Fluid dynamic model, river meandering model, numerical simulation, logistic model	River dynamics "induce vegetation patterns"	Hypothetical (using data from SC, USA, Pearlstine and others. 1985)
Stand characteristics	Stromberg and others 1993	Depth to groundwater	Depth to ground-water	Stand biomass (leaf area index and vegetation volume), stand structure (maximum canopy height, and basal area), leaflet variables (primary leaflet area, primary leaflet length, and secondary leaflet number), xylem water potential	Second order linear regression		Hassayampa R., San Pedro R., Tanque Verde Ck., AZ, USA
Cover types/ Functional groupings	Toner and Keddy 1997	Depth, duration, and timing of flooding, fraction of the growing season flooded, last day of first flood, length of the second flood, mean depth of flooding, number of floods per growing season, number of days of drawdown preceding midseason floods, time of second flood	Depth, duration, and timing of flooding	Presence-absence of woody cover	Logistic regression	Static channel geometry	Ottowa R. Ontario, Canada
Cover types/ Functional groupings	Richter and Richter 2000	Duration of flooding above threshold (effective discharge)	Duration of flooding above threshold (effective discharge)	Abundance of patch types	Numerical model simulations	Flood driven meandering drives forest succession; channel maintenance approach	Yampa R., CO, USA

USDA Forest Service Gen. Tech. Rep. RMRS-GTR-282. 2012

91

Chapter 6: Case Studies

Case Study I: Mount Emmons iron fen, Crested Butte, Colorado

Introduction and problem statement

The expansion of hard rock mining for molybdenum on Mount Emmons could alter the watershed, which supports a unique iron fen located at the mountain base. In contrast to bogs, which are rain and snowmelt driven, fens are supported primarily by groundwater (refer to Chapter 1). The fen is a State of Colorado Natural Area and supports a continuous carpet of acid loving *Sphagnum* mosses as well as the regionally rare carnivorous plant, *Drosera rotundfolia* (round leaf sundew). Two rare species of dragonfly (*Leucorhinea hudsonica* and *Sematochlora semicircularis*) are also found in the fen. The project goal was to understand the hydrologic and geochemical regime that supports the fen.

Background

Many fens in the Colorado Rocky Mountains are supported by groundwater discharge from hillslope aquifers. The Mount Emmons fen, located west of the town of Crested Butte in the Grand Mesa, Gunnison, and Uncompahgre National Forest, is geochemically distinct because its watershed contains pyrite-rich bedrock. The oxidation of pyrite produces sulfide, which in solution, produces sulfuric acid that leaches metal ions from rock resulting in acidic and heavy metal-rich water (Cooper and others 2002). These ecosystems are termed "iron fens" because dissolved iron is transported to the fen where it oxidizes and precipitates onto organic particles to form limonite, or bog iron ore deposits. Iron fens are the only wetland ecosystem type in the region with water that has a high natural acidity. The combination of peat and limonite accumulation creates landforms unique to iron fens, including pools and metal rich terraces (Figure 6-1). In addition, iron fens support rare plant species and unique plant communities.

Figure 6.1. Mount Emmons iron fen, Crested Butte, Colorado.

USDA Forest Service Gen. Tech. Rep. RMRS-GTR-282. 2012

93

Approach

This study focused on quantifying the seasonal variation in groundwater levels, piezometric heads, water chemistry, and vegetation in the fen while trying to identify the water source(s) supporting the fen. Six water table monitoring wells, 1 staff gauge in the pond, and 10 piezometers were installed, all into peat soils. Instruments were installed in nests consisting of one water table well and two or three piezometers with openings at different depths. The location of instrumented sites is shown on Figure 6-2. Each water table well and the staff gauge had a continuous water level recorder. After installation, well casing lengths and stick up height were measured. Field measures of water pH were made by bailing water from monitoring wells. In addition, water samples were collected for laboratory chemical analysis of major cation and anion concentrations. Topographic surveying was used to record the fen boundary and the elevation, latitude, and longitude of all wells, staff gauges and other topographic features. Daily precipitation was analyzed with a logging rain gauge. Correlation analysis was used to relate water levels in monitoring wells to each other.

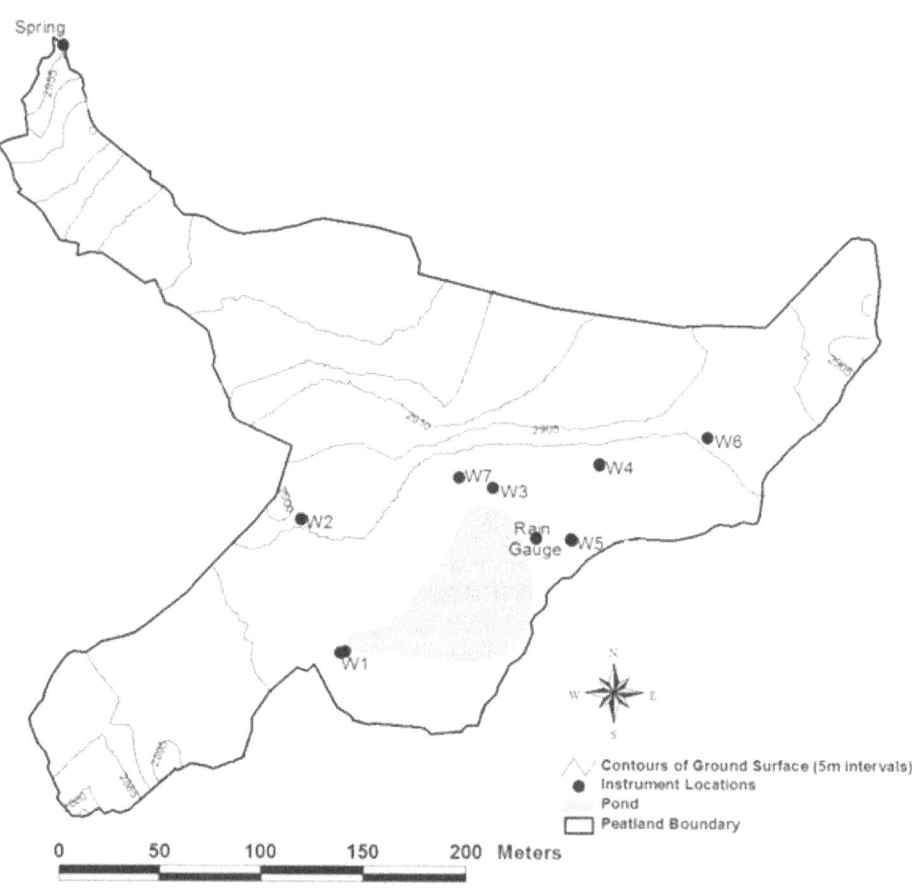

Figure 6-2. Topographic map of Mount Emmons fen. Well 1 (W1), a staff in the fen pond (gray area), is located near the pond outlet. Monitoring wells 2 through 7 are also identified. Contour interval is 5 m. The fen is 15.1 acres (6.1 ha) in size.

Results

Winter precipitation was averaged over the 2002 and 2003 seasons. June and early July 2003 were dry, but strong monsoonal flow produced regular rain from late July through mid-September. The pond water level varied little during the study period. However, groundwater levels in water table wells varied considerably (Figures 6-3 and 6-4).

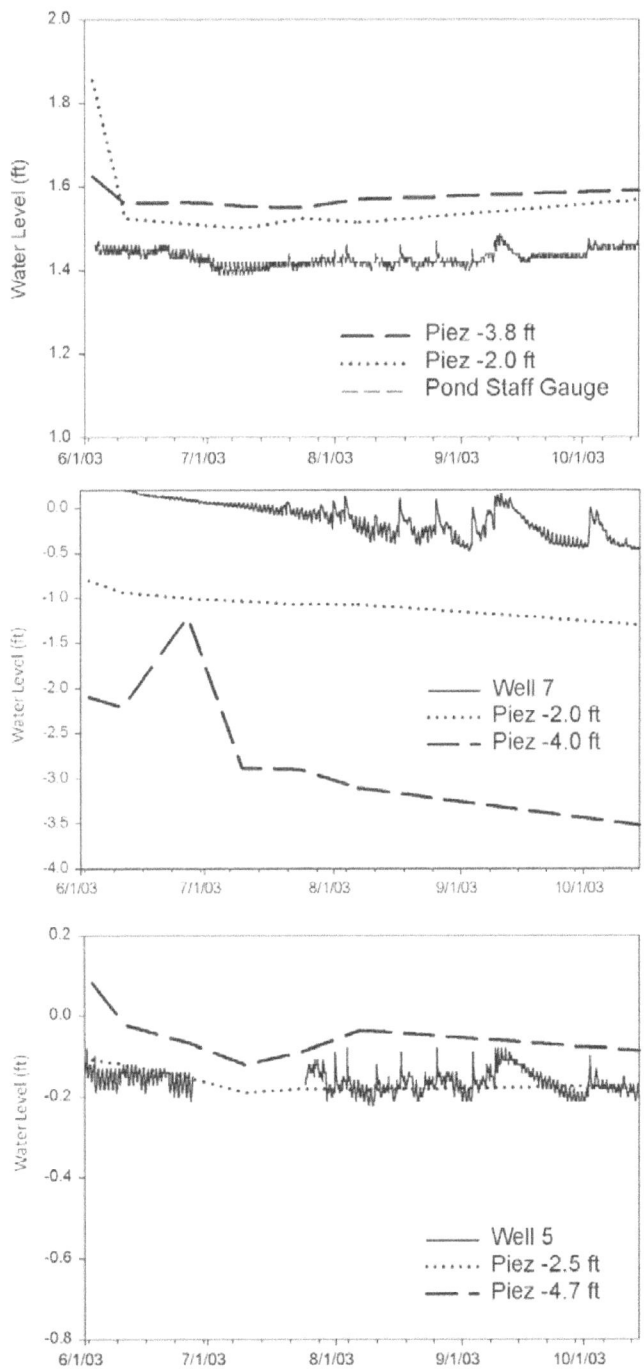

Figure 6-3. Water levels of staff in pond (top) and depth to groundwater (ft) in wells 7 (middle) and 5 (bottom) relative to piezometers. The figure legend indicates the depth at which each piezometer is completed.

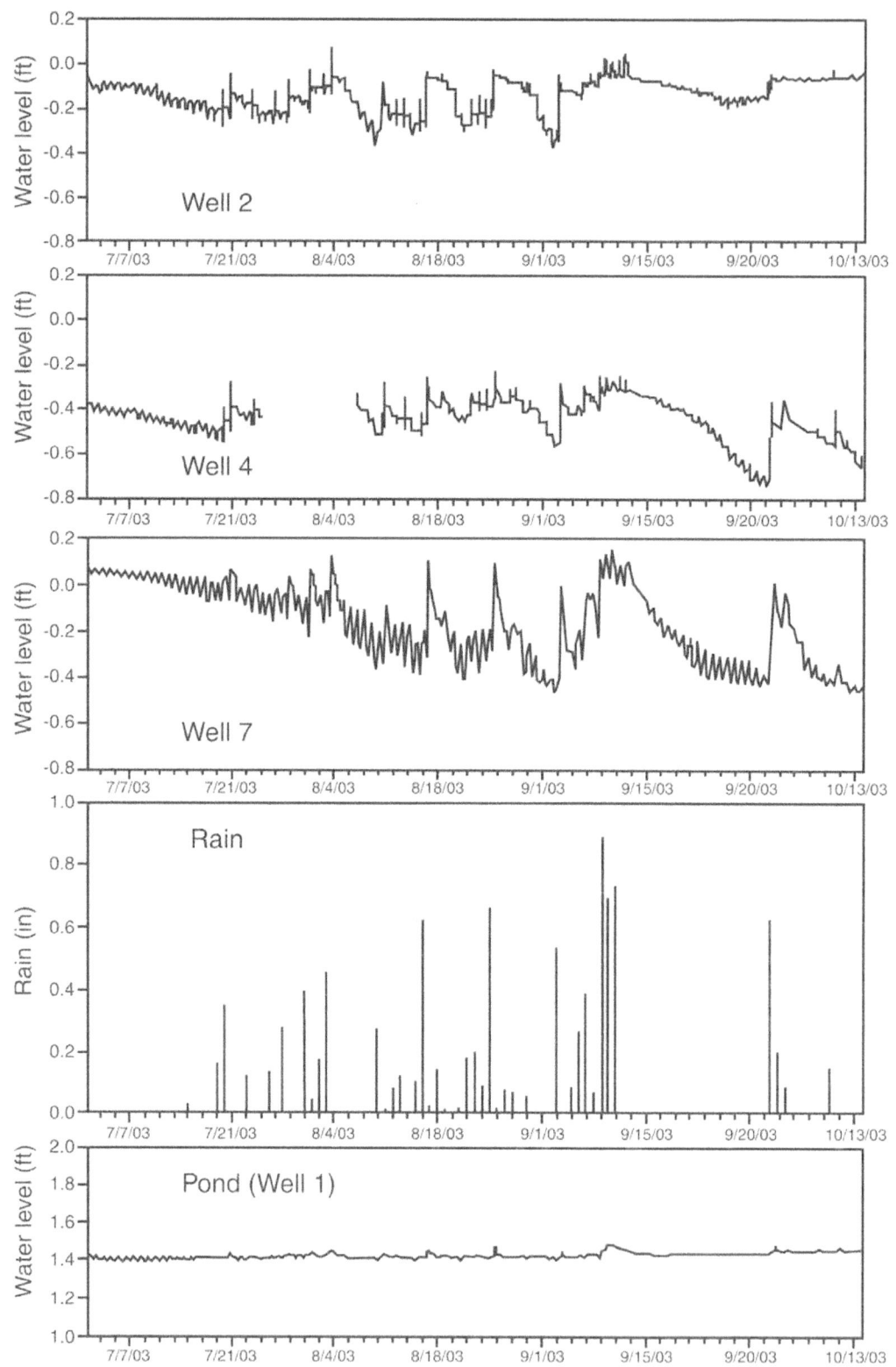

Figure 6-4. Groundwater response at wells 4 and 7 and pond (well 1) and daily precipitation totals for summer 2003.

Most wells had water levels near the soil surface in early summer 2003 due to snowmelt recharge of hillslope aquifers. Water levels in wells 2 through 6 remained near the soil surface for most of the summer of 2003, while water levels at well 7 declined during the summer. Measured head in piezometers was higher than water levels in monitoring wells at all sites except well 7, indicating upward vertical gradients (Figure 6-3) and groundwater discharge to the soil surface at most sites, including under the pond. In addition, groundwater discharging to the surface maintained sheet flow through portions of the site. This water had its origin along the northern fen edge. All wells and the pond level responded to summer precipitation events, rising after rains and declining during rainless periods. This was apparent in September 2003 when a several-week rainless period caused groundwater levels to fall (Figure 6-4).

Vegetation of the Mount Emmons fen is dominated by acid-tolerant plant species, primarily *Sphagnum angustifolium, S. russowii, S. fimbriatum, Carex aquatilis, Carex utriculata, Betula glandulosa, Drosera rotundifolia, Vaccinium scoparium, Calamagrostis canadensis, Pinus contorta,* and *Picea engelmannii* (vascular plant species nomenclature follows Weber and Wittmann [2001], and bryophyte nomenclature follows Crum and Anderson [1981]). The western portion of the fen is a water track with sheet flowing water and supports a monoculture of *Carex aquatilis.* The area just west of the pond and extending north through the center of the fen has an open overstory of *Picea engemannii* and *Pinus contorta* with an understory of *Carex aquatilis* or *Calamagrostis canadensis.* The water track on the eastern side of the fen, which flows into the pond, has numerous small, unvegetated pools with vegetated strings between the pools. *Carex aquatilis* is the main sedge in these areas, and small populations of *Drosera rotundifolia* occur, especially near wells 3 and 5.

Surface and groundwater flow was from north to south at all times during the study period due to the steep topographic and hydrologic gradient that controls surface and groundwater flow direction (Figure 6-5). Well 1, the pond staff, was most closely correlated with well 2, which occurs west of the pond, but it was poorly correlated with other wells (Table 6-1). Well 4 was highly correlated with wells 3, 5, 6, and 7.

Surface water and shallow groundwater sampled from water table wells were highly acidic (Figure 6-6), particularly in well 6. Groundwater discharged and flowed south down the limonite water track and into the pond. The pond water (well 1) had a pH similar to wells 4 and 5. The three most westerly wells—2, 3, and 7—had higher pH than other waters during the fall of 2002. Wells 2 and 7 had similar pH values on 1 July 2003, while the pH of well 3 dropped to 3.9. The water in piezometers had pH greater than 6.0, even beneath the pond or shallow groundwater (Figure 6-7). Thus, two waters of distinct pH and likely chemical content meet within the fen, and yet it is the acid water of the springs fed by the Mount Emmons pyrites that controls the surface water and shallow groundwater chemistry and that facilitates the presence of acid loving and acid tolerating biota as well as the formation of limonite landforms.

These data suggest that the northwest portion of the fen received the most acid water and that acid concentration decreased by an order of magnitude in the groundwater flow system to the west (near well 2) and by two to three orders of magnitude in the water beneath the fen. Thus, the acidity of water varied spatially, and acid concentration varied by approximately 1000 times between water in well 6 and water in the deepest piezometer at well 6. The water at all sites was dominated by calcium and sulfate ions. Sulfate likely originated from the oxidation of pyrite (FeS). There was no detectable HCO_3 or NO_3 in any water sample.

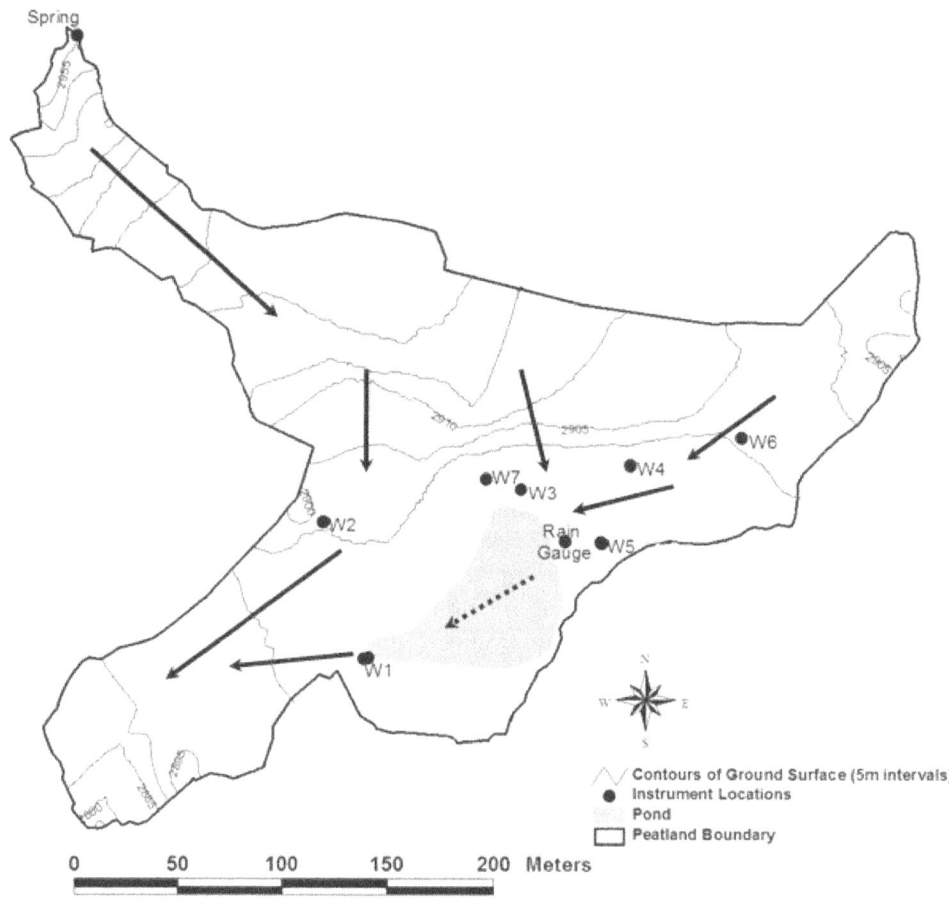

Figure 6-5. Topographic map (contours are 5 m) of Mount Emmons fen with arrows showing principal surface and groundwater flow directions.

Table 6-1. Correlation coefficients (r) between monitoring wells and pond for 2003. Each row and column represents a well, and input data are depth to water table on each measurement date.

	2	3	4	5	6	7
1	0.558	-0.069	0.086	0.268	0.187	0.131
2		0.022	0.167	0.258	0.348	0.281
3			0.654	0.339	0.608	0.283
4				0.647	0.763	0.614
5					0.426	0.467
6						0.637

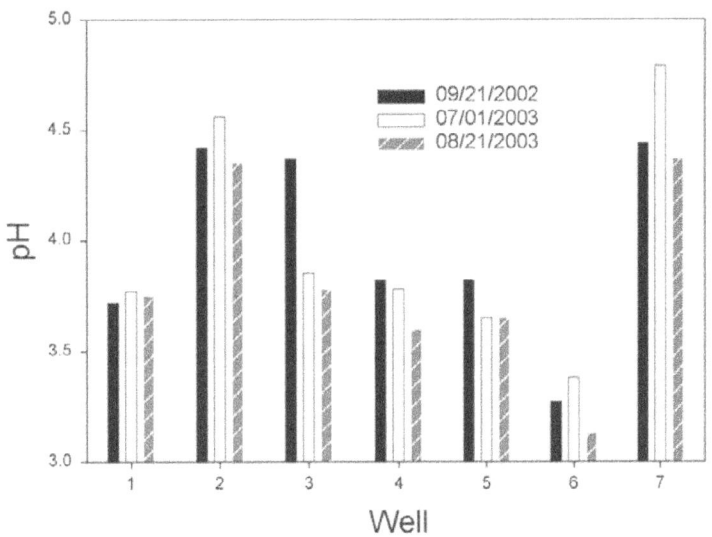

Figure 6-6. Surface water pH at pond (well 1) and six wells (wells 2-7), Mount Emmons fen.

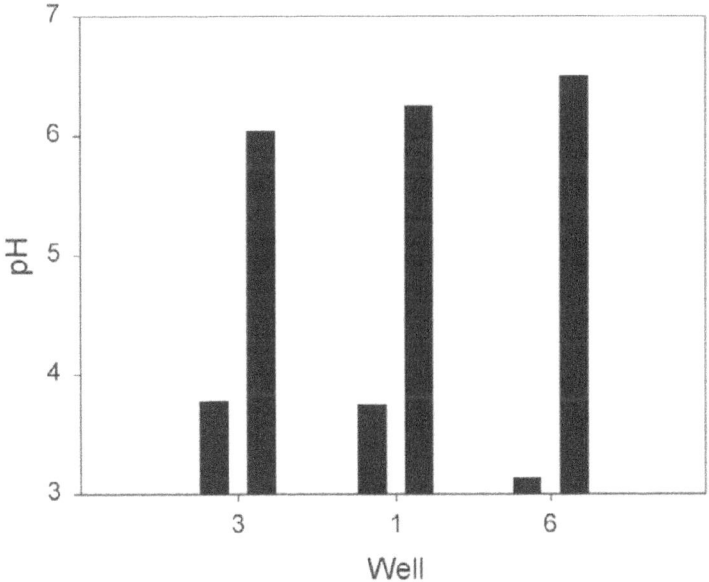

Figure 6-7. pH values for surface water or shallow groundwater (lower pH value to left) and piezometer water (higher pH value to right) at wells 3, 1, and 6.

USDA Forest Service Gen. Tech. Rep. RMRS-GTR-282. 2012

99

Discussion

Groundwater discharging from the base of Mount Emmons produces surface sheet flow and groundwater flow that perennially saturates the fen. Water levels remained relatively stable at most sites, suggesting that a large volume of groundwater flows through the fen. Because the fen has been accumulating peat for more than 8000 years (Fall 1997), it is likely that hydrologic processes have been relatively intact for a long time. Late summer rain recharges hillslope aquifers, resulting in short-term increases in groundwater levels. The fen's watershed is large and complex with both short and long flow paths, which is indicated by perennial groundwater flowing through the fen during the summer of 2002, even under extreme drought conditions.

Perennial groundwater inflow is critical to drive hydrologic and geochemical processes that lead to peat formation. In addition, Mount Emmons fen is geochemically and ecologically unique. The acidic water is produced from oxidizing pyrite and veins of pyrrhotite, another iron sulfide mineral, which is common in areas of contact metamorphism. These rocks surround a molybdenum ore body on Mount Emmons. Acidic water produces the unique hydrologic and geochemical environment and an ecological refugium for *Drosera rotundifolia*, *Sphagnum angustifolium*, *Betula glandulosa*, and other acid-loving species. These species are dependent upon the perennial flow of acidic water. It is critical to understand that groundwater with pH greater than 6.0 is present beneath the fen and has sufficient hydraulic head to reach the peat body. If the flow of acid water was reduced, the alkaline water could dominate fen geochemistry, resulting in the loss of acid-tolerant taxa with no potential for replenishment (Cooper and others 2002).

Key methodological approaches used in this study include the use of water table well and piezometer nests and analyzing water for pH and chemical content. The pond water level is controlled by groundwater inflow and sheet flow, but its level varies little over time and is not representative of the natural variation in groundwater inflow within the fen. Long-term protection of this fen will require maintaining groundwater inflows to maintain groundwater levels throughout the fen, and the water must have the natural range of acidity that has been measured. Groundwater levels should be at or above the soil surface in early summer. Water levels may drop during the summer, although rarely by more than about 50 cm below the soil surface. Groundwater levels typically rise in the mid- to late summer due to monsoon rains.

Case Study II: Groundwater pumping in a riparian ecosystem, South Platte Park, Littleton, Colorado

Introduction and problem statement

Groundwater can be pumped from alluvial aquifers for agricultural, industrial, or municipal uses. In some cases, groundwater is pumped directly from beneath riparian zones, resulting in a cone of depression and larger-scale lowering of the water table. Groundwater pumping is commonly used to avoid surface water diversions on sand bedded streams and where the construction of water diversion structures is difficult.

Most phreatophytes are groundwater dependent and have high water demand during hot summer days. Some are highly sensitive to even small changes in water availability. For example, *Populus deltoides* (plains cottonwood) is among the most sensitive tree in North America to drought-induced water stress and xylem embolism formation (Tyree and others 1994). Embolisms are air bubbles that form in vessels when xylem pressure potential lowers to the point where chains of water molecules break and gaseous air blocks the movement of water up the vessel. Embolisms make vessels dysfunctional for water transport. If enough vessels develop embolisms, the plant will lose a significant proportion of its water transport ability and will suffer leaf or twig dieback or whole plant death. While groundwater pumping is generally increasing across the West, we still know relatively little about the effects of water table declines on most plant species.

The purpose of this case study was to investigate the effects of groundwater pumping from under the South Platte River floodplain south of Denver, Colorado, on plains cottonwood trees. The Centennial Water and Sanitation District (Centennial) had water rights along the South Platte River. Instead of building a diversion structure, officials chose to install a groundwater pumping system. This portion of the river is within South Platte Park, which is managed by the South Suburban Parks and Recreation District and the City of Littleton, Colorado. For this project to be permitted by the City of Littleton, it was necessary for Centennial to demonstrate that groundwater pumping would not harm or kill the cottonwoods along the river.

In the southern portion of this reach, the South Platte River had been mined for gravel approximately 30 years prior to this study and had been influenced by a large flood in 1973. During this flood, the river channel avulsed to the west and the old channel filled with sand and gravel, allowing a new cohort of cottonwoods to establish. The former mined area had sand and gravel soils, relicts of the floodplain soils. The natural floodplain area had never been mined and had fine-textured soils in the upper 1 m. Large populations of cottonwood trees were present on both the mined and un-mined sites, providing numerous trees of similar age. The four pumping wells that had been installed by Centennial were located in a transect that fortuitously spanned the area between the mined and un-mined river reaches (Figure 6-8). Each well was capable of pumping at a sustained rate of approximately 30,000 m³/day.

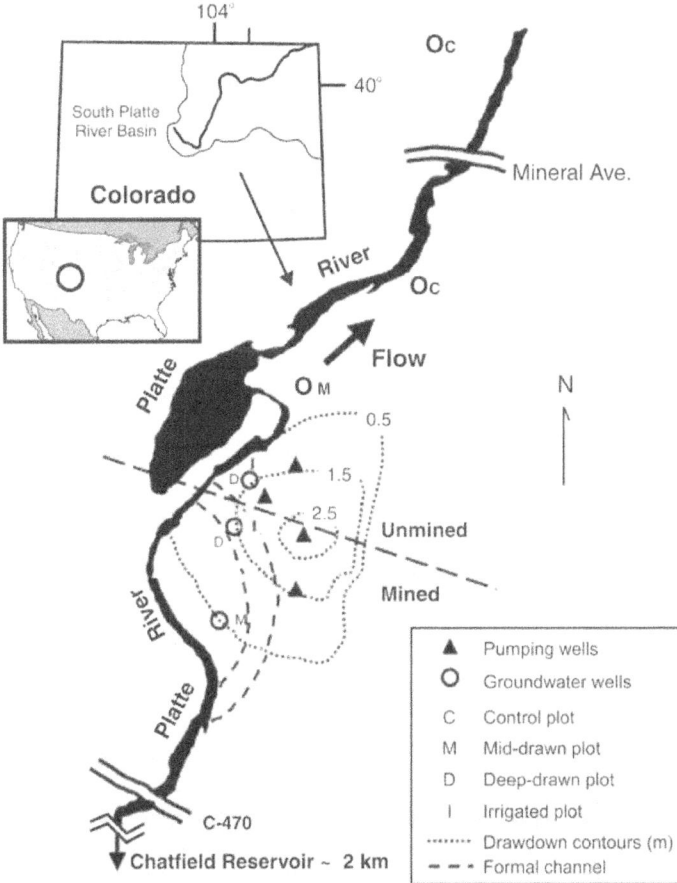

Figure 6-8. Study area location showing drawdown contours, pumping wells, and plot locations.

USDA Forest Service Gen. Tech. Rep. RMRS-GTR-282. 2012

101

Methods

To determine the potential effects of groundwater pumping on the water table configuration, a 30-day pump test was performed in January of 1996, and water table drawdown was measured in 18 monitoring wells placed through the study area. The drawdown extent and depth (cone of depression) data were used to create a water table drawdown map (Figure 6-8). Drawdown was as much as 2.5 m. Based on these initial findings, a study was designed to analyze the effects of drawdown amount in each soil type (gravel in the mined area and natural soils in the un-mined area) on the riparian cottonwoods. Using the water table drawdown map, six study plots, each 600 m² in area and with similar tree density (1000 trees/ha), were chosen to represent deep drawdown (1.5 to 2.0 m), moderate drawdown (0.3 to 0.6 m), and no drawdown (controls) in the mined and un-mined substrate areas. Each plot was equipped with a water table monitoring well, and water table depth was measured manually at least weekly during the study period. One additional deep drawdown plot was established in each soil type to test the effects of supplemental water addition (2.54 cm every three days) applied by sprinklers as a watering treatment.

Groundwater pumping began on 3 July and ended on 27 July 1996. The depth to the water table was measured weekly prior to and following pumping through the entire summer, and daily during the pumping period for the six wells in the study plots.

Five cottonwoods that represented the range of tree size in each plot were analyzed for their physiological response to the change in water table depth. Xylem pressure potential (Ψ_{xp}) was measured on one full sun terminal twig on each study tree using a Scholander-type pressure bomb during pre-dawn (00:00 to 05:00) and midday (12:00 to 15:00) periods. Sampling occurred weekly during the pre-pumping period in the summer of 1996, and three times each week during the pumping period. Two observers estimated percentage of leaf yellowing and senescence on 25 randomly selected trees in each plot during the pumping period. Stomatal conductance (g_s) was measured using a null balance porometer (LiCor LI-1600) on hourly time steps approximately three weeks after the pumping ceased to evaluate plant recovery patterns.

The pre-dawn and midday xylem pressure potential data from the six main plots were analyzed using multivariate repeated measures analysis of variance (MANOVA) with the GLM procedure in SAS (Cooper and others 2003b). The study design followed a split plot design with soil type, drawdown depth, and time as the independent variables. Three time periods were analyzed: (1) pre-pumping, (2) the first three sample dates in the pumping period, and (3) the last three sample dates in the pumping period. Two-way ANOVA was used to test the effect of watering on predawn xylem potential in time period (3). Regression analysis was used to model the relationship between maximum plot water table drawdown and pre-dawn water potentials on 22 July 1996 as well as the percent of leaves that yellowed.

Results

The water table depth was 1.5 to 2.5 m belowground in the pre-pumping period. South Platte River flows increased in late June and mid-July and caused a rise in the water table, particularly in the mined study reach. Groundwater pumping produced a large drawdown in the deep drawdown plots and a smaller drawdown in the mid drawdown plots, as predicted (Figure 6-9). When pumping ceased, all water tables rose rapidly.

Midday xylem pressure potential (Ψ_{md}) declined during the pre-pumping period as summer temperatures increased (Figure 6-10). Pre-dawn xylem pressure potentials (Ψ_{pd}) declined very slowly during the pre-pumping period and changed little in the first two weeks of the pumping period. There were no differences in Ψ_{md} or Ψ_{pd} among plots in the pre-pumping period. On 15 July, Ψ_{pd} in the deep and mid drawdown plots declined

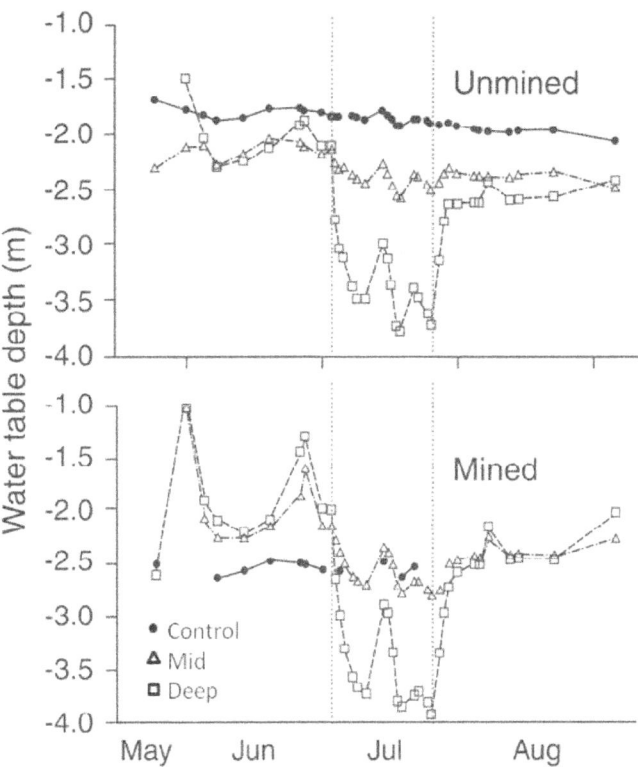

Figure 6-9. Depth to water table for all study plots during the pre-pumping, pumping, and post-pumping periods.

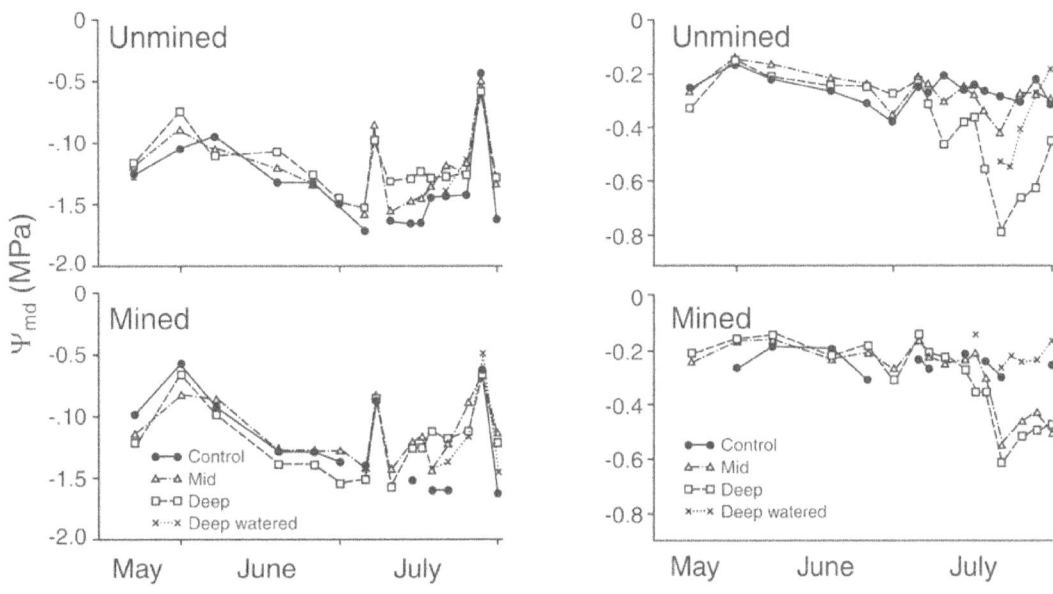

Figure 6-10. Midday (left panel) and pre-dawn (right panel) xylem pressure potentials for the for all study plots during the pre-pumping, pumping, and post-pumping periods.

sharply, while in the control plots and the watered mined plot, Ψ_{pd} remained similar to the pre-pumping period. The lowest Ψ_{pd} occurred on 22 July, and many cottonwood leaves in the affected plots turned yellow and eventually fell from the trees due to the development of extreme water stress and cavitation of most xylem in twigs. Once leaf yellowing occurred, Ψ_{pd} rose due to the reduced leaf area and water demand.

Depth of water table drawdown and percent leaf loss were significantly correlated with tree Ψ_{pd} on 22 July (Figure 6-11). The depth of water table drawdown was linearly related to the decline in tree Ψ_{pd} and the loss of leaves. At the end of the pumping period, Ψ_{pd} was higher in watered than un-watered plots at the mined site, but Ψ_{pd} was intermediate between what was measured in deep and mid drawdown plots in the unmined site. The drawdown reduced soil water availability, which lowered xylem pressure potential, caused the development of embolisms, and led to leaf and twig dieback and loss. At the end of the pumping period, Ψ_{pd} was lower in watered plots, especially in the mined plot. In addition, the two watered plots had much lower leaf loss than trees in the deep or mid drawdown plots.

Three weeks after the cessation of pumping, Ψ_{xp} measured hourly was similar for watered and un-watered deep drawdown plots (Figure 6-12). However, trees in the watered plots reached a higher g_s and maintained higher stomatal conductance for longer time during the day than did un-watered plots. Thus, watered plots maintained a higher level of stomatal conductance and, most likely, a higher rate of photosynthesis compared with un-watered plots.

Discussion

Groundwater pumping and other hydrologic changes to riparian and wetland ecosystems can produce both short-term and long-term effects on plants, soils, and ecological processes. This experiment demonstrated the short-term effects that can result from lowering the water table under riparian cottonwoods. *Populus deltoides* is highly sensitive to reduced water availability; within two weeks, Ψ_{pd} declined rapidly, leading to leaf senescence, twig death, and some whole branch sacrifice (death). The loss was limited because pumping was ceased as soon as these effects were visible or measureable on the trees. The response of trees to the water table decline was not immediate,

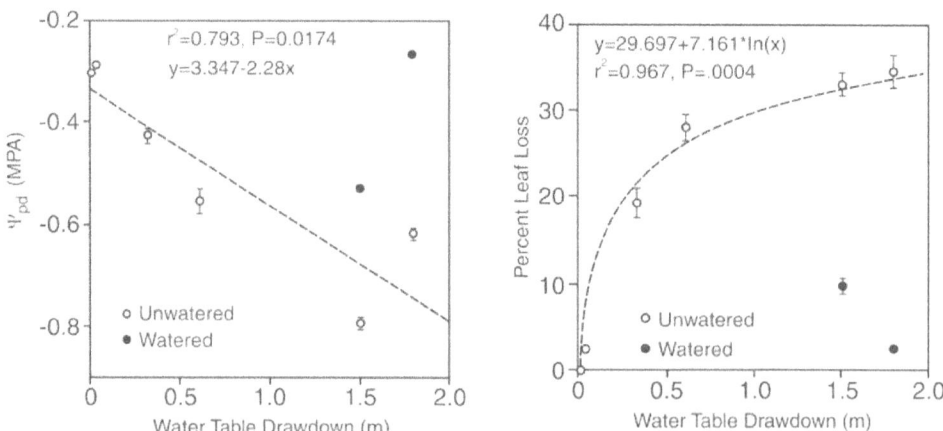

Figure 6-11. Affects of water table drawdown on pre-dawn water potential (left panel) and percent leaf loss (right panel) for all study plots.

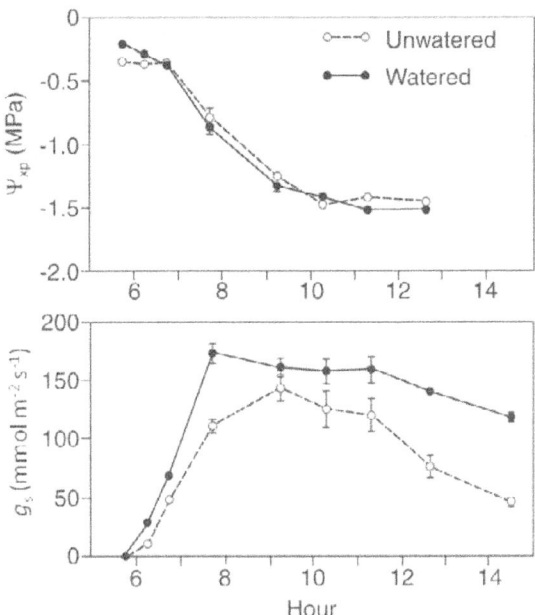

Figure 6-12. Hourly midday water potential (top panel) and gs (bottom panel) for watered and unwatered deep drawdown plots.

likely due to a short increase in South Platte River discharge, which raised river stage and groundwater levels as well as soil water storage, which provided suitable water for trees for a short period. While runaway xylem cavitation, or embolism formation, has been reported in eastern populations of cottonwood, this study demonstrated that under natural conditions trees experienced low water potentials (Ψ_{md} readings of -1.8 MPa) but were able to recover their water potentials each night. Thus, western populations of this tree species are more drought tolerant than eastern populations. Xylem cavitation appeared to be caused by the reduction in Ψ_{pd}, indicating that plants could no longer acquire sufficient water from the soil to recover each night.

Leaf loss was related to the magnitude of water table drawdown. Even a relatively modest water table drawdown of 0.3 m decoupled some or most of the tree roots from their water source, resulting in leaf death. A deeper drawdown of greater than 1.0 m caused significant whole tree stress. Had this drawdown been sustained, many trees would have died, as has been reported for other areas (Scott and others 1999).

Application of a relatively small amount of irrigation water allowed trees to maintain their water status with little or no apparent cavitation and leaf loss in study area trees, even where deep drawdown occurred. This suggests that trees can effectively switch from primarily using groundwater to water in upper soil horizons. The effect was more obvious in the sand and gravel mined soil study area than in the fine-grained, un-mined soil area, likely due to the coarse soils allowing water to percolate more rapidly to plant roots and because these soils supported fewer herbaceous plants.

Case Study III: Marsh hydrology, Great Sand Dunes National Park, Colorado

Introduction and problem statement

Marshes form in basins where seasonal surface water and/or groundwater create periodic deep standing water that supports highly productive plant communities. Marshes are among the most important wetlands for supporting migratory birds and amphibians. However, marshes are highly sensitive to the volume of surface water or groundwater delivery that fills basins each year, and even small changes in water delivery can lead to their drying.

Between 1937 and 1995, a complex of more than 100 marshes disappeared from a sand dune complex at Great Sand Dunes National Park in Colorado (Figures 6-13 and 6-14). These changes were documented using vertical air photographs, but the cause was unknown. Three overall hypotheses were advanced: (1) sand dune migration during a regional drought in the 1950s buried the marshes; (2) regional groundwater pumping lowered the water table, leading to their drying; and (3) changes in local hydrologic processes led to wetland loss (Wurster and others 2003). These three possibilities were investigated using stream flow records, groundwater level measures, natural stable isotopes as a tracer of water sources, buried soil seed banks as an indicator of where wetland horizons existed, and groundwater modeling.

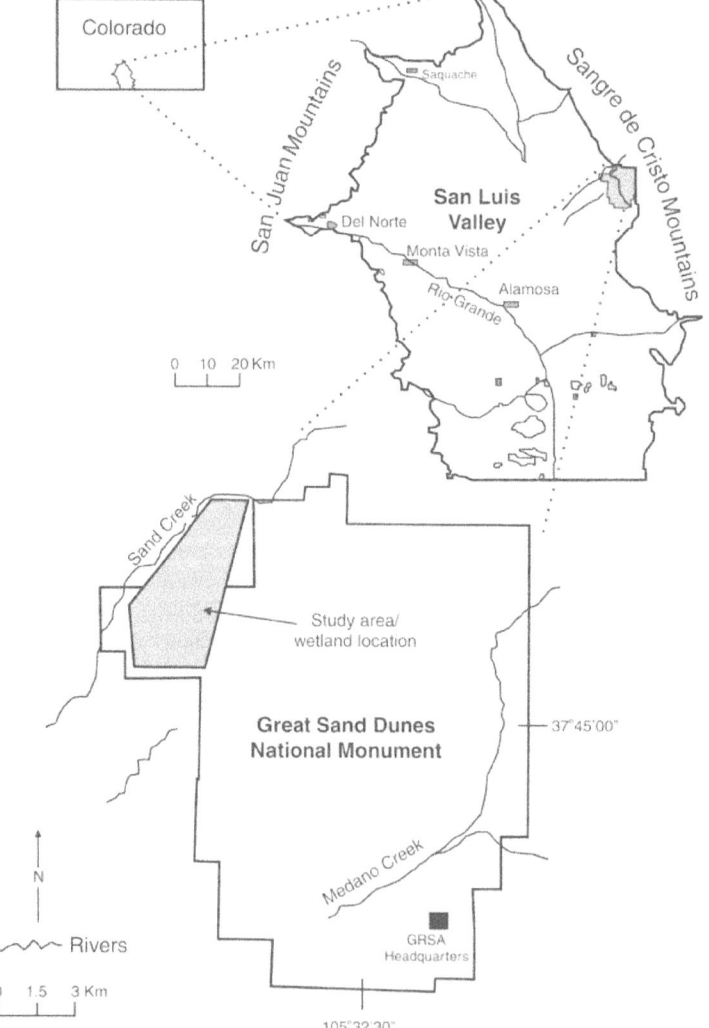

Figure 6-13. Location of study area along Sand Creek, Great Sand Dunes National Monument, Colorado.

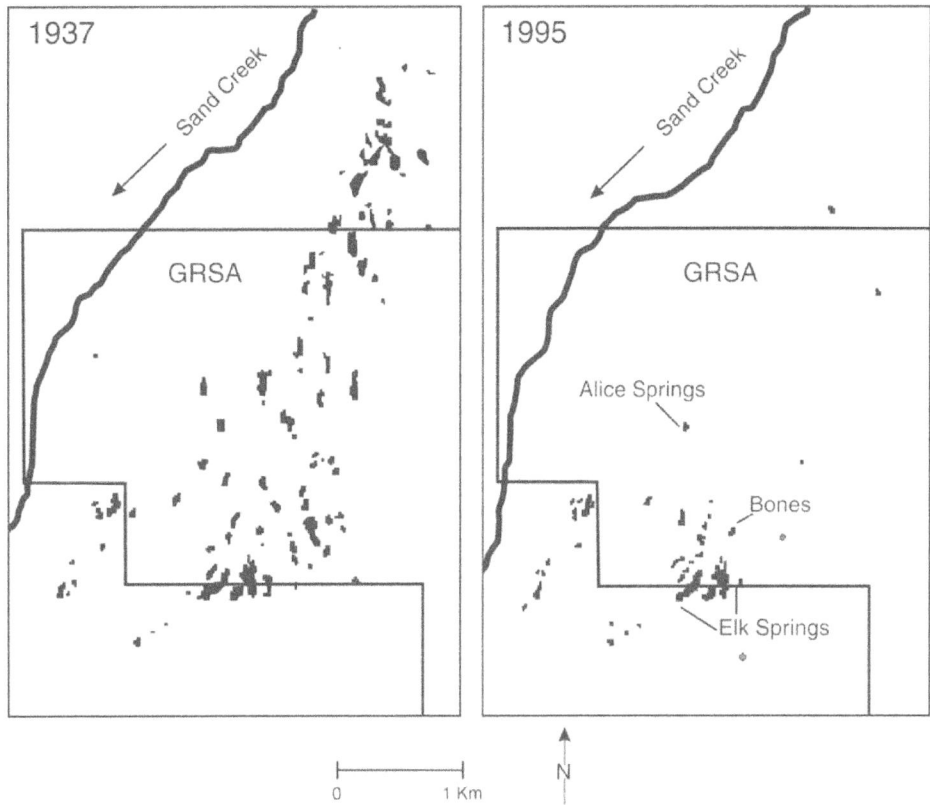

Figure 6-14. Distribution of marshes, black polygons, in 1937 and 1995 showing extensive loss of the northern marshes.

Methods

The studied wetland complex occurred within a sand dune field. There was little surface runoff due to the high permeability of the sand. Instead, the marshes were fed by groundwater. To investigate the groundwater source(s) and seasonal variation in water levels, a total of 120 water table wells were installed by hand in and around both the marsh complexes that existed during the study period as well as those identified in the 1930s air photos, and in transects that extended across Sand Creek (Figure 6-15). Staff gauges were installed in Sand Creek as well as in marshes with standing water. Aquifer transmissivity (T) was estimated using the stage ratio and time lag methods (Ferris 1963). Average T was determined by calculating the time lag between maximum water level fluctuations at different wells in the study area. Specific yield (Sy) was not directly measured—a value of 0.15 was used. Since the aquifer was nearly homogeneous sand, little variance in its estimate would have occurred.

Stable isotope ratios of δD and $\delta^{18}O$ in water samples were used to identify recharge and flow paths for the marshes. Monthly water samples were collected from rain events and surface water in Sand Creek. Groundwater samples were collected from 10 water table wells along transect A-A' and B-B' (Figure 6-15). Samples were collected using standard protocols, stored in airtight vials, and frozen until analyzed; and all wells were bailed prior to sample collection. Analysis was conducted using a VG Optima mass spectrometer.

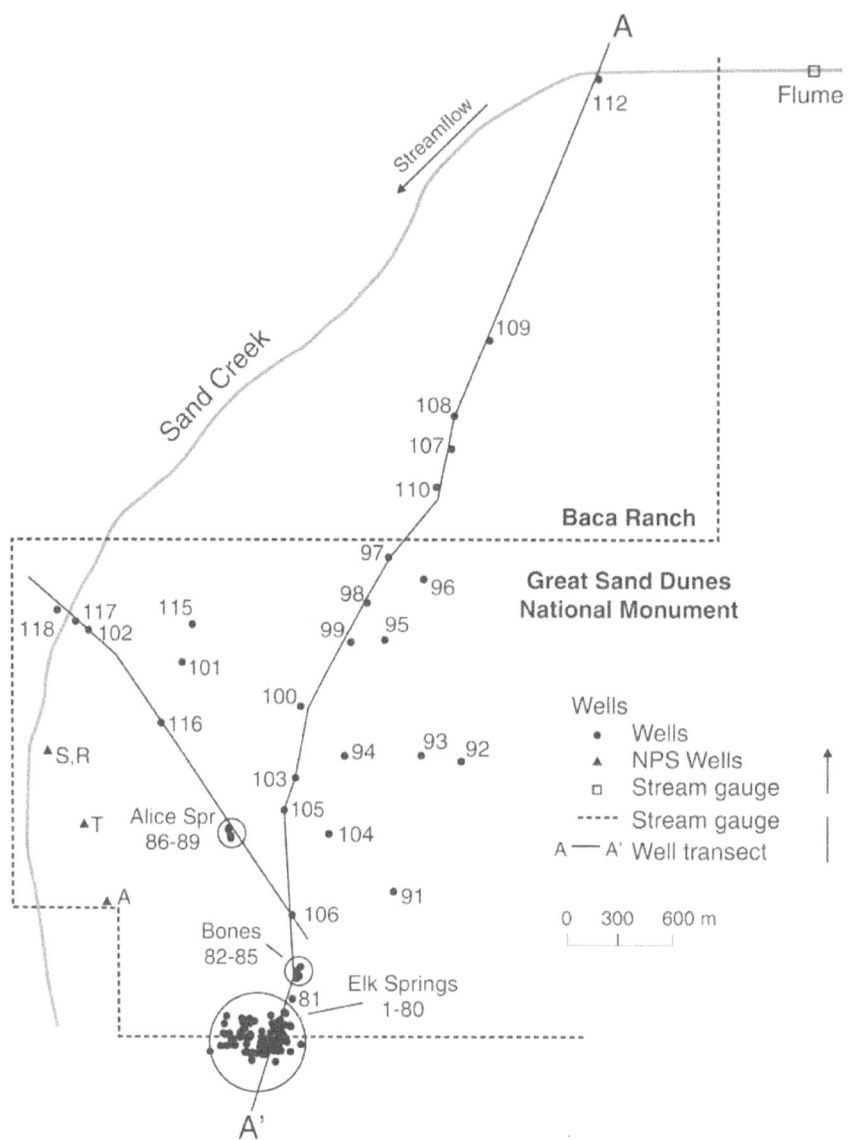

Figure 6-15. Location of monitoring wells.

A key goal of this analysis was to understand how the unconfined aquifer responded to changes in flow of Sand Creek, a potential recharge source for the marshes. Seasonal fluctuations in groundwater levels were modeled by comparing water level change in monitoring wells under Sand Creek and monitoring wells at various distances away from Sand Creek (Wurster and others 2003).

Historical monthly Palmer Drought Severity Index (PDSI) values for the study area were obtained from the National Climate Data Center. PDSI can be used to assess the severity of wet and dry weather patterns (Palmer 1965). Historical records of mean daily discharge for several regional streams with long-term records were used to reconstruct the flow of Sand Creek, which had discharge data available only from 1994 to 1999. All streams were within 20 km of Sand Creek, had similar watershed areas and lithologies, and were at similar elevations. From these data, a rough flow-duration curve was constructed for different years and was used to estimate the aquifer response to stream flow from 1936 to 1999. The aquifer's response as measured by the water table elevation was calculated at a point 1500 m from Sand Creek.

A 6-km reach of Sand Creek's channel was analyzed using historical air photographs to determine whether channel form changes occurred that could have influenced groundwater recharge patterns or processes. The former locations of marshes were identified in the field using GPS points, and soils were investigated to identify wetland horizons with organic accumulations, gleying or mottling, and the depth of burial, if any. Soil samples from any buried wetland horizons found during sampling were collected and analyzed for percent organic matter by loss on ignition. The soil seed bank was analyzed by spreading soils onto trays in a greenhouse where they were watered to maintain saturation for six months. Emerging seedlings were identified and counted.

Results

Mottled soil horizons were interpreted as identifying the seasonal water table high and were used to identify the former wetland soil surface. This surface was buried more than 0.5 m belowground at a few sites, near wells 97, 98, 99. At all other sites, the wetland horizons were at the ground surface. Thus, sand burial of the wetland surface was not common in the study area. However, the water table measured during this study was well below these mottled horizons, indicating that the wetland soils formed when the water table was at a higher elevation.

The plant species that germinated from the wetland soils included: *Eleocharis palustris* (spike rush), *Scirpus pungens* (three square bulrush), and *Juncus arcticus* (arctic or Baltic rush), all of which are common and dominant wetland plant species in the study area. The presence of germinable seeds indicated that the soil horizons investigated supported wetland plants during the Twentieth Century and were the wetlands visible on the 1930s era air photographs. Percent organic matter was highest in horizons with mottles and that contained germinable seed, providing multiple lines of evidence that these horizons were wetland soils.

Water table contours indicated that a regional hydraulic gradient exists between Sand Creek and the study wetlands (Figure 6-16). The maps presented illustrate conditions when Sand Creek was not flowing (17 Feb 1999) and was flowing (15 July 1999). Water table elevation is highest along the Creek and in the northeastern portion of the study area. Therefore, the dominant groundwater flow through the sand sheet was from the northeast to the southwest.

Groundwater profiles along transects A-A' (Figure 6-17A) further clarified the strong groundwater gradient from Sand Creek toward Elk Springs. The profile at B-B' (Figure 6-17B) indicated that a substantial groundwater mound forms under Sand Creek when it flows, and the mound dissipates during the summer and fall. When Sand Creek is flowing, the water table is in contact with the channel bed. Once surface flow ceases in this intermittent stream, the water table declines rapidly. The high soil permeability (hydraulic conductivity of approximately 10^{-4} m/s) facilitates groundwater flow from the channel area into the sand sheet and toward the wetland basins.

USDA Forest Service Gen. Tech. Rep. RMRS-GTR-282. 2012

109

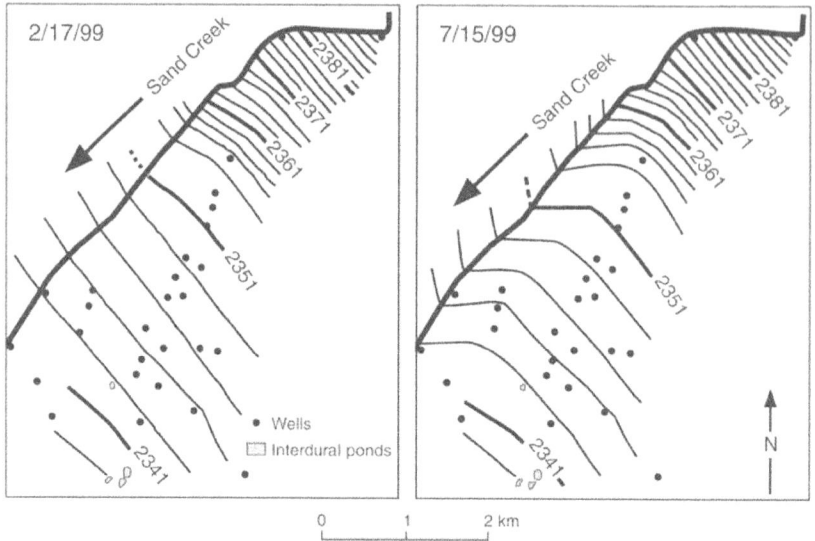

Figure 6-16. Water table contours in the study area for February (left) and July (right) 1999 when the creek is not and is flowing, respectively. Note contour lines straight across creek in February and bending downstream in July. This indicates that a groundwater mound forms under the creek when it flows.

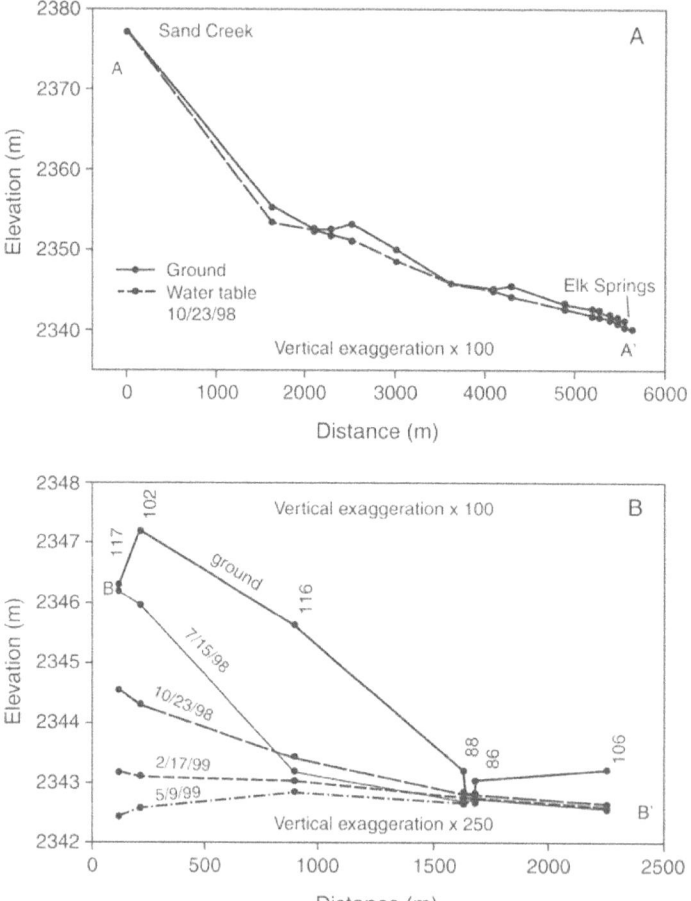

Figure 6-17. Profiles A-A' and B-B' showing the elevation of the water table along a nearly 2500 m long transect on different dates in 1998 and 1999.

Hydrographs from four wells (101, 100, 94, and 93) at increasing distances from Sand Creek depict water table maxima increasing in time from September through March (Figure 6-18). A graph of the time lag in days of the groundwater maxima in monitoring wells following Sand Creek's peak flow indicates a nearly linear temporal increase with distance from the creek (Figure 6-19). These data suggest that Sand Creek surface flows produce a pulse of groundwater recharge that propagates as a wave through the sand sheet and supports groundwater flow through the wetlands. Isotopic analysis of the $\delta D/\delta^{18}O$ ratio of creek and groundwater indicated that Sand Creek is the source of water for the groundwater flow system that supports the wetlands.

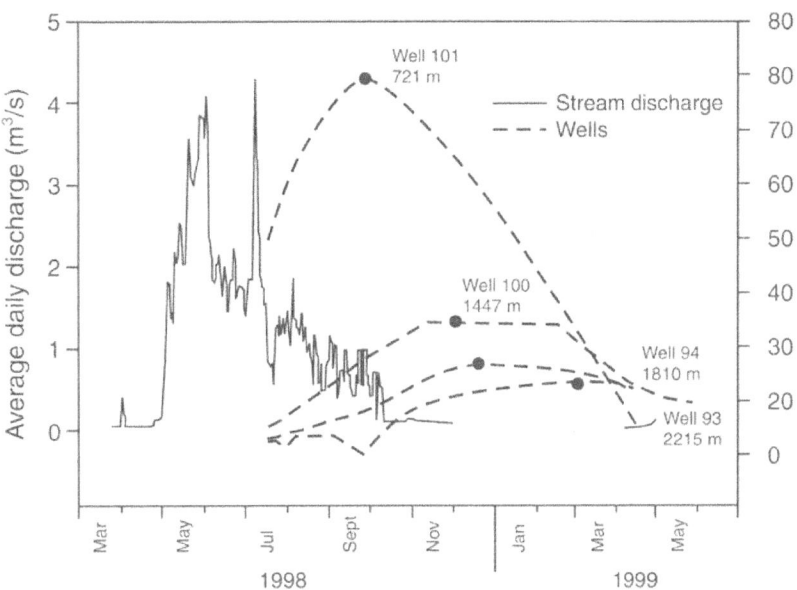

Figure 6-18. Sand Creek average daily discharge during 1998 and groundwater levels for four wells during 1998 and 1999.

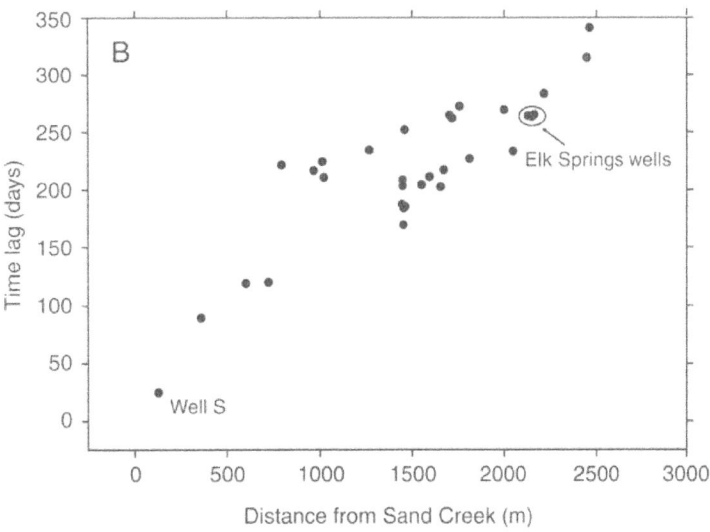

Figure 6-19. Time lag of maximum groundwater high relative to distance from Sand Creek.

PDSI values for the Twentieth Century indicated that severe drought occurred during the 1930s and 1950s and that wetter periods occurred during the 1910s, 1940s, 1980s, and 1990s. Analysis of the Sand Creek channel indicated that it narrowed approximately 50 percent between 1937 and 1995 and that sinuosity also increased during the same period. The relict channel boundaries from 1937 could be identified in the field using indicators such as large woody debris, cobbles, stream terraces, and cottonwood establishment sites. The difference in elevation of the channel at the time of analysis (1999) in 1937 was 2.5 m, suggesting that the channel incised over the last 60-year period, leading to channel narrowing and lowering of the groundwater base level.

Discussion

These analyses provided a wide ranging and complex set of studies, from which it was concluded that burial by moving dunes could explain only a small fraction of the wetland loss. There was no evidence that regional groundwater pumping was connected to wetland loss; however, local hydrologic processes could explain most of it. The study marshes are fed by groundwater that is recharged during high flow in Sand Creek as a result of early summer snowmelt from the Sangre de Cristo Mountains. Surface water recharges the groundwater table, building an elevated groundwater mound under Sand Creek that dissipates as it flows away from the creek. The groundwater flow produced by the dissipating mound takes 200 to 300 days to reach many of the wetland basins.

The soil morphology, organic matter, and soil seed bank analyses all indicated that the former wetland soil surface could be identified and was near the current ground surface. But the current groundwater table was well below the ground surface. A groundwater model that projected the water table for a point 1500 m from Sand Creek using the stream elevation in 1937 and in 1995 after 2.5 m of incision occurred indicated that the water table in 1995 would be greater than 0.5 m below the level seen on 1937 air photographs and that stream incision is the likely cause of wetland loss (Figure 6-20).

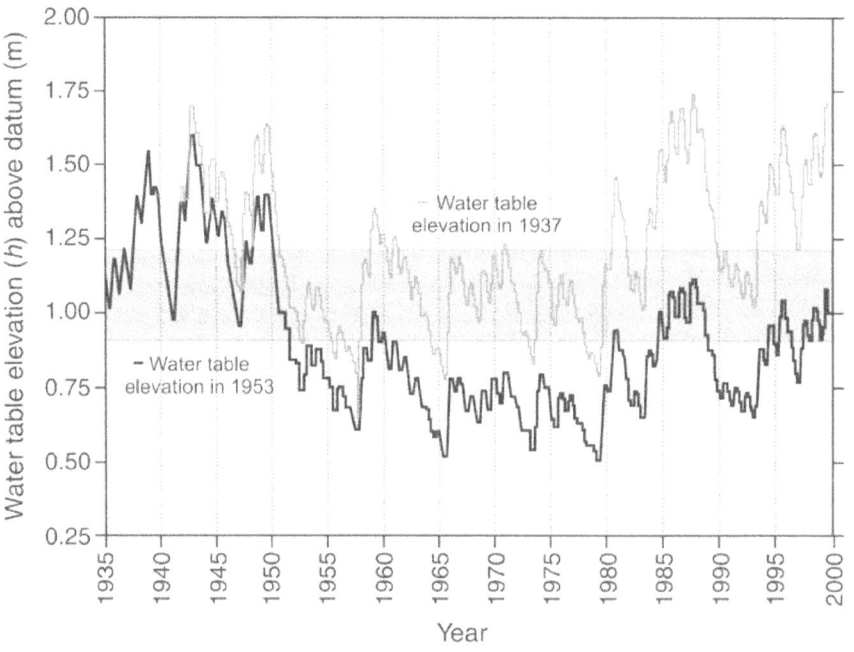

Figure 6-20. Groundwater elevation for a point 1500 m from Sand Creek modeled for the period 1937 to 2000 using 1937 stream bed elevation (light line) and current stream bed elevation (dark line).

Even though years of high snowpack and Sand Creek runoff occurred during the 1980s and 1990s, the water table under the sand sheet has not elevated enough to flood the wetland basins. The ultimate cause of stream channel change is unknown. It could have been caused by cottonwood establishment along the floodplain margin during a period of fluvial stability, hydrologically suitable conditions, or fire suppression in the San Luis Valley. The trees could have stabilized banks, constrained the channel, and caused it to shift from braided to meandering form. The concentration of flow in the stable narrower channel may have resulted in incision. Other possible causes are heavy livestock grazing in the channel area or several particularly large flow events.

Case Study IV: Instream flow needs to support groundwater and riparian vegetation, Cherry Creek, Arizona

Introduction and problem statement

The U.S. Forest Service filed for an instream flow water right in 1999 for Cherry Creek, which is a perennial stream on the Tonto National Forest in central Arizona. The State of Arizona approved the requested flows (median monthly), which the agency filed for to support wildlife habitat, fishery, and recreation. To better understand the relationships among surface flow in Cherry Creek, groundwater across the valley bottom, and the riparian vegetation, a study was conducted along Cherry Creek (Merritt and others 2010b). Conservation of riparian habitat along the creek is important because riparian habitats constitute less than 1 percent of the entire landscape, yet they provide a variety of unique resources, functions, recreational opportunities, and habitat qualities not found in adjacent upland habitats.

Cherry Creek flows from spring-fed canyon reaches into a wide alluvial valley bottom through which the channel loses to the alluvial aquifer flow for most of the year (Figures 6-21A and B). The stream supports diverse deciduous riparian forest that is comprised of trees such as *Populus fremontii* (Fremont cottonwood), *Salix gooddingii* (Goodding's willow), *Jugulans major* (Arizona walnut), *Fraxinus velutina* (velvet ash), and *Alnus oblongifolia* (Arizona alder).

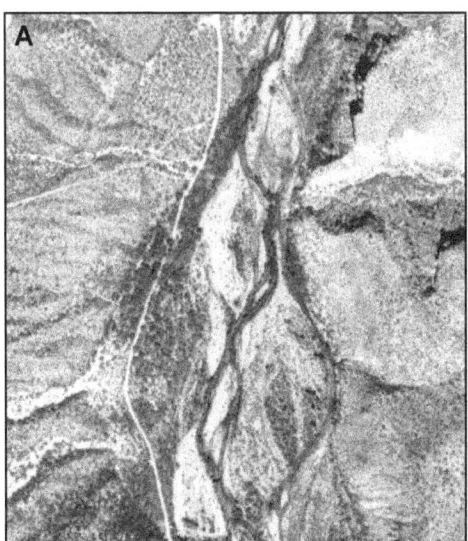

Figure 6-21. (A) Aerial photograph of the study reach of Cherry Creek, Tonto National Forest, Arizona. (B) Cherry Creek looking downstream.

USDA Forest Service Gen. Tech. Rep. RMRS-GTR-282. 2012

113

Methods

Ten water table wells and two well-staff gauges (wells positioned immediately adjacent to the stream) were established along a 0.5-km alluvial reach of Cherry Creek. The wells were established in a grid covering much of the valley bottom on both sides of Cherry Creek. Elevation at each of the wells was surveyed, and wells were instrumented with Onset Hobo data loggers set to record water depth every 15 minutes. One well was fitted with a non-submerged pressure transducer and data logger for later barometric correction of water levels. Groundwater surfaces were generated by spline fitting a surface through measured groundwater depths (Figure 6-22). Vegetation cover types were classified from vegetation sampling plots distributed along transects throughout the study area. Cover types were modeled as a function of depth to water table at a stable low flow along Cherry Creek using discriminant function analysis. Groundwater decline was simulated throughout the study area, and levels were used to predict changes in vegetation cover types in the study reach.

A physical model for Fremont cottonwood distribution as a function of groundwater depth was constructed using the topography-bathymetry coverage, the water table at low flow, and measurements of the mean of six cottonwood root depths from excavations conducted in the study area. Potential cottonwood habitat (groundwater within rooting zone) was modeled as a function of manually simulating a lowering of the water table. Detailed topography (±0.01m) and channel bathymetry through the reach was available from an associated USGS study (Waddle and Bovee 2009).

Linkages between groundwater elevations and fluctuations and streamflow in Cherry Creek were determined through examining hydrographs (stage and discharge) of flood pulses and response of groundwater levels at wells across the floodplain, correlating daily groundwater fluctuations with fluctuations in discharge at a USGS gauge 15 km upstream from the study area, and developing rating curves (stage-discharge relations) at the staff gauges.

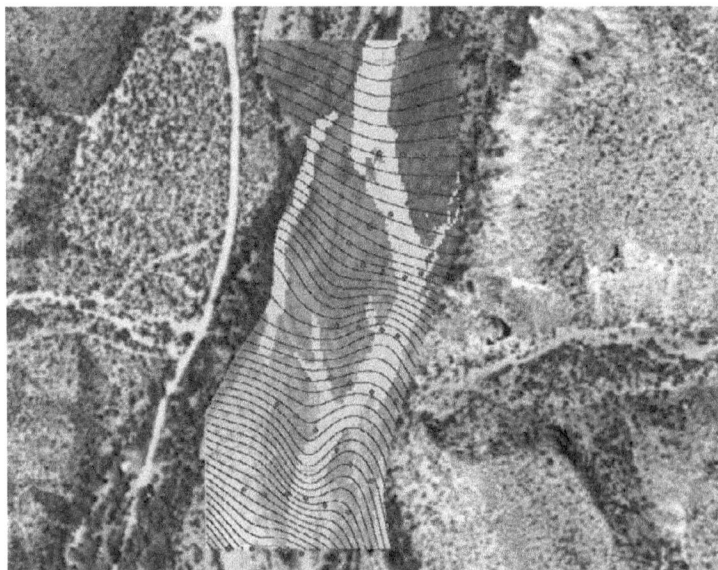

Figure 6-22. Contour map of water table (purple surface) with 0.1 m groundwater contours, water table monitoring wells (blue cross) and vegetation plots (red symbol), and channel (green). Streamflow is from top to bottom of frame. The groundwater flow is general trending down valley and toward river right (left side of frame). Scale is 1:3000.

Streamflow in Cherry Creek was related to water status of important riparian tree species by measuring pre-dawn and midday water potential using PMS Model 670 Scholander-type pressure chambers (PMS Instruments, Corvallis, Oregon) in individuals located upstream and downstream from a diversion that withdraws 95 percent of the flow in Cherry Creek. The relationship between tree fitness (growth rate) and discharge was also examined by coring cottonwood and willow, measuring ring widths, and relating these widths to specific monthly flow attributes.

Results and conclusions

Riparian tree species were most abundant in areas along Cherry Creek where groundwater was within 2 m of the ground surface, which corresponds reasonably well with the average rooting depth measured from excavated roots (1.74 m). Statistical models and physical simulation models predicted that incremental decimeter reductions in surface water and groundwater levels to 1 m below base level would cause declines in cottonwood-dominated riparian forest from a current level of 41 percent of sites sampled to 7 percent at an average loss rate of 3 percent per decimeter decline in groundwater level from the modeled low flow (Figure 6-23). The model also predicted that shrub-dominated habitat would increase from 59 percent to 93 percent with a 1 m decline in groundwater levels. At a 2-m decline in groundwater, only 1.8 percent of the riparian habitat would be habitable by cottonwood, and the valley bottom would be comprised exclusively of shrubland (98.2 percent frequency).

These findings were independently supported by simulation models that determined potential cottonwood habitat through modeling the area of the valley bottom with groundwater within the average rooting depth of cottonwood over a range of groundwater levels. Potential cottonwood habitat declined 75 percent (from 65 percent to 16 percent of the valley bottom) with a simulated 1-m decline in groundwater level (Figure 6-23). Potential cottonwood forest was virtually eliminated at 1.5- and 2.0-m reductions in groundwater below the modeled low flow (reduced to 7 and 3 percent potential habitat, respectively).

The models also illustrated that groundwater decline would result in migration of the riparian forest-xeroriparian edge nearer the stream channel and a reduction in the extent, or complete loss, of riparian forest, depending on the severity and persistence of flow-induced groundwater decline.

Reductions in forest cover and tree species diversity were documented along a perennial to intermittent reach caused by a flow diversion (95 percent of flow diverted) as well as an additional perennial to intermittent reach where surface water is controlled by underlying bedrock. Forest species cover dropped from 62 to 10 percent in comparisons of perennial to intermittent reaches, and shrubs comprised the dominant cover type in the intermittent reach. Downstream from the diversion, riparian tree cover in the vegetation plots averaged 4.5 percent compared to 23 percent in forest cover upstream.

Flow diversion caused increases in xylem water stress of riparian cottonwood, Arizona sycamore, and Goodding willow, but saltcedar individuals were not critically affected by flow diversion (refer to Figure 5-6 in Chapter 5). The consequences of surface water depletion and its effect on water availability to trees included: diminished cover of native riparian trees, and higher cover of xerophytic (drought-tolerant) desert shrubs (including non-native saltcedar) along the flow-diverted reach.

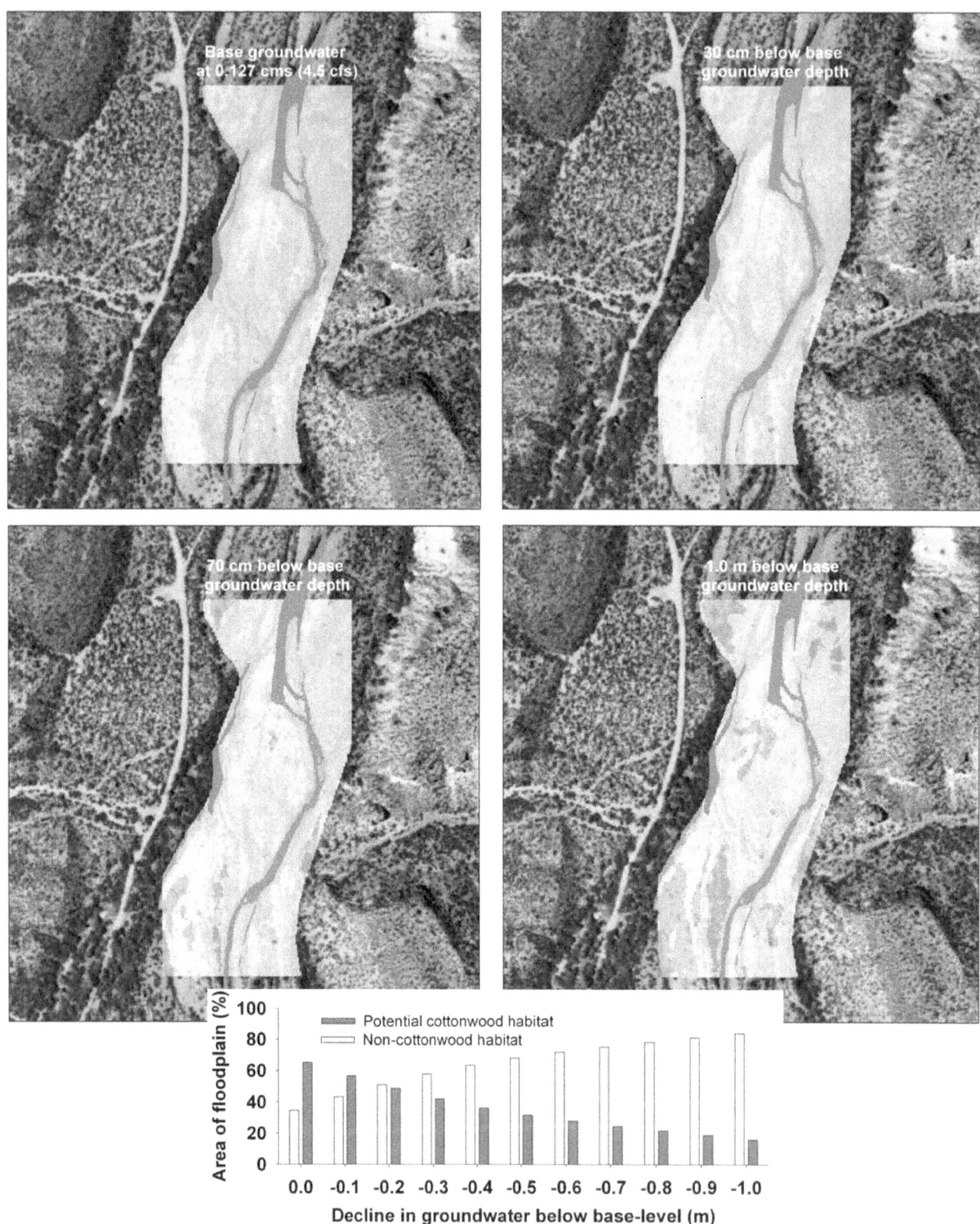

Figure 6-23. Potential cottonwood habitat (green) and unsuitable cottonwood habitat (yellow, orange, and red). Frames show potential habitat at a base groundwater level (at discharge in Cherry Creek of 0.127 cms [4.5 cfs]), and at 30, 70, and 100 cm drawdowns. Area of potential habitat declines from 65 to 16 percent of the modeled floodplain area as a function of a 1-m drawdown below base level.

USDA Forest Service Gen. Tech. Rep. RMRS-GTR-282. 2012

Variability in growth rates of cottonwood and Goodding willow could be explained by differences in streamflow from year to year (Figure 6-24). Mean May, June, and July annual streamflow rates each explained most of the variability in tree growth rates. The median monthly instream flows filed for by the Forest Service would support intermediate growth rates of cottonwood and Goodding willow. Incremental reductions in flow would result in decreased cottonwood growth rates of 1.5, 3, and 3.5 mm per 5 cfs reduction in the months of May, June, and July. Similar reductions in Goodding willow growth were found. Cottonwood and Goodding willow are obligate phreatophytes; therefore, their health and persistence depend largely on maintenance of water tables within reach of their rooting zone. These linkages between streamflow and growth rates provided evidence that groundwater through the study reach is fed by streamflow. Analysis of groundwater levels across the valley and comparison of these fluctuations with surface water fluctuations and those measured at a streamflow gauge 15 km upstream further evidence a direct connection between groundwater levels and streamflow.

These findings suggest that the species comprising riparian forests along Cherry Creek depend on alluvial groundwater systems that are supported by and dependent upon streamflow in Cherry Creek. Maintenance of sufficient streamflow to support shallow water tables reduces the likelihood of conversion of cottonwood and Arizona sycamore-dominated forests to shrub communities and those more typical of upland and xeroriparian habitats. Reductions in streamflow and subsequent groundwater decline would suppress riparian forest species and favor non-native saltcedar and desert shrubs, resulting in a loss of riparian habitat. If the vegetation along Cherry Creek shifted from a cottonwood-willow-dominated riparian forest habitat to a saltcedar-burrobrush shrubland habitat, it would likely support mostly upland species found in adjacent xeroriparian areas instead of both riparian and upland species.

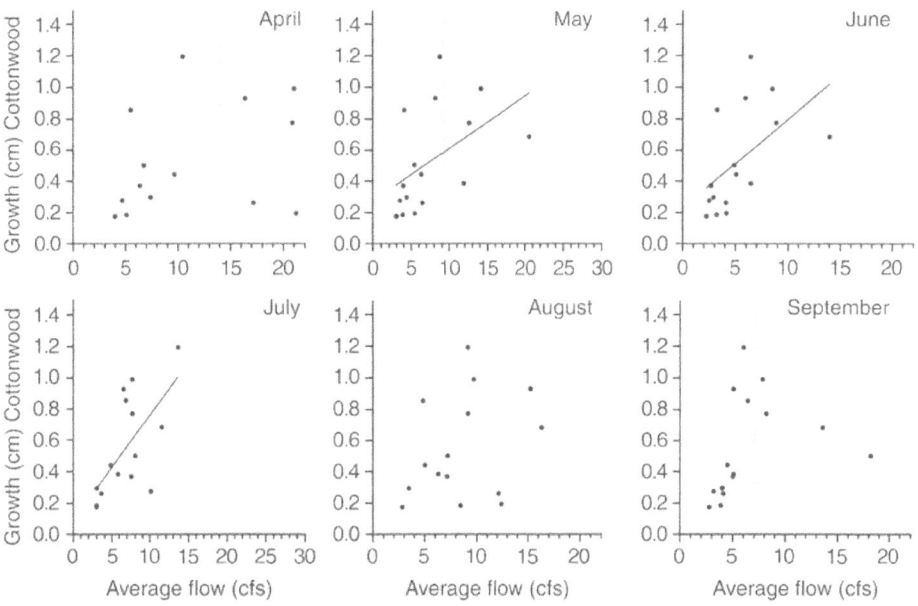

Figure 6-24. Relationships between average flow (Q in cfs) calculated for each month during the growing season and related to cottonwood growth for each year. Only significant relationships are plotted. May: Growth (mm) = 0.28 + 0.033Q; r^2 = 0.25; P = 0.05. June: Growth (mm) = 0.23 + 0.056Q; r^2 = 0. 28; P = 0.03. July: Growth (mm) = 0.09 + 0.067Q; r^2 = 0.43; P = 0.008. Relationships between April, August, and September flow and incremental growth were not significant (P>0.05). Vertical dotted lines indicate the median instream flow for each month (period of record).

Incremental reductions in streamflow would have negative consequences for riparian forest species and associated wildlife habitats including: (1) elevated water stress and decreased growth rates (fitness) of riparian tree species, (2) reduction in spatial extent of potential habitat for dominant forest species (e.g., cottonwood), and (3) likely conversion of complex riparian forest cover types to structurally and compositionally simple (or homogeneous) shrubland and bare cover types.

Acknowledgments

We thank Joyce Van De Water for assistance with the figures. Review and comments from Christopher Carlson, Joseph Gurrieri, Kristen Kaczynski, Rob Hubbard, John Potyondy, and Lindsay Reynolds improved this report. We also thank Vernon Bevan and Camille Febvre for critical review and for proofing the text.

Literature Cited

Adair, E. C., D. Binkley, and D. C. Andersen. 2004. Patterns of nitrogen accumulation and cycling in riparian floodplain ecosystems along the Green and Yampa Rivers. Oecologia 139: 108-116.

Alberta Agriculture and Food. 2004. Salinity Classification, Mapping and Management in Alberta. Alberta Agriculture and Agriculture Canada, Alberta, Canada. http://www1.agric.gov.ab.ca/$department/deptdocs.nsf/all/sag3267.

Alstad, K. P., S. C. Hart, J. L. Horton and T. E. Kolb. 2008. Application of tree-ring isotopic analyses to reconstruct historical water use of riparian trees. Ecological Applications 18:421-437.

Arscott, D. B., K. Tockner, and J. V. Ward. 2003. Spatio-temporal patterns of benthic invertebrates along the continuum of a braided alpine river. Archiv Fur Hydrobiologie 158:431-460.

Auble, G. T., J. M. Friedman, and M. L. Scott. 1994. Relating riparian vegetation to present and future streamflows. Ecological Applications 4:544-554.

Auble, G. T., and M. L. Scott. 1998. Fluvial disturbance patches and cottonwood recruitment along the upper Missouri River, Montana. Wetlands 18:546-556.

Auble, G. T., M. L. Scott, and J. M. Friedman. 2005. Use of individualistic streamflow-vegetation relations along the Fremont River, Utah, USA to assess impacts of flow alteration on wetland and riparian areas. Wetlands 25:143-154.

Baker, W. L. 1990. Climatic and hydrologic effects on the regeneration of *Populus angustifolia* James along the Animas River, Colorado. Journal of Biogeography 17:59-73.

Baron, J. S., N. L. Poff, P. L. Angermeier, C. N. Dahm, P. H. Gleick, N. G. Hairston, R. B. Jackson, C. A. Johnston, B. D. Richter, and A. D. Steinman. 2002. Meeting ecological and societal needs for freshwater. Ecological Applications 12:1247-1260

Bedford, B. L., and K. S. Godwin. 2003. Fens of the United States: distribution, characteristics, and scientific connection versus legal isolation. Wetlands 23:608-629.

Bio/West, Inc. 1986. Final Report--Environmental evaluation for Sandhills Cooperative River basin study. Bio/West, Inc., Logan, Utah, USA.

Birken, A. S., and D. J. Cooper. 2006. Processes of *Tamarix* invasion and floodplain development along the lower Green River, Utah. Ecological Applications 16:1103-1120.

Blake, T. J., J. Sperry, T. Tschaplinski and S. Wang. 1996. Water relations. Pages 401-422 *in* R. Stettler, H. Bradshaw, P. Heilman and T. Hinkley, editors. Biology of *Populus* and its implications for management and conservation. NRC Research Press, National Research Council of Canada, Ottawa, Ontario.

Bohn, H.L. 1971. Redox potentials. Soil Science 112:39-45.

Bowden, W. B. 1987. The biogeochemistry of nitrogen in freshwater wetlands. Biogeochemistry 4:313-348.

Brinson, M. M. 1993. A hydrogeomorphic classification for wetlands. Technical Report WRP-DE-4. U.S. Army Engineer Waterways Experiment Station, Vicksburg, Mississippi, USA.

Brinson, M. M., and A. I. Malvarez. 2002. Temperate freshwater wetlands: types, status, and threats. Environmental Conservation 29:115-133.

Brown, D. J., W. A. Hubert, and S. H. Anderson. 1996. Beaver ponds create wetland habitat for birds in mountains of southeastern Wyoming. Wetlands 16:127-133

Brown, T.C. 2004. The marginal economic value of streamflow from National Forests. Discussion Paper DP-04-1, USDA Forest Service Proceedings RMRS-4851. Rocky Mountain Research Station, Fort Collins, Colorado, USA.

Buchanan, T. J., and W. P. Somers. 1976. Techniques of water-resources investigations of the United States Geological Survey. Applications of Hydraulics, Chapter A8, U.S. Geological Survey, Washington, DC, USA. Second edition.

Bush, S.E., K.R. Hultine, J.S. Sperry, and J.R. Ehleringer. 2010. Calibration of thermal dissipation sap flow probes for ring- and diffuse-porous trees. Tree Physiology 30:1545-1555.

Busch, D. E., N. L. Ingraham, and S. D. Smith. 1992. Water uptake in woody riparian phreatophytes of the southwestern United States: a stable isotope study. Ecological Applications 2:450-459.

Busch, D. E. and S. D. Smith. 1995. Mechanisms associated with decline of woody species in riparian ecosystems of the southwestern U.S. Ecological Monographs 65:347-370.

Camporeale, C., and L. Ridolfi. 2006. Riparian vegetation distribution induced by river flow variability: a stochastic approach. Water Resources Research 42: W10415, doi:10.1029/2006WR004933.

Carsey, K., D. Cooper, K. Decker, D. Culver, and G. Kittel. 2003. Statewide wetlands classification and characterization—Final Report. Colorado Department of Natural Resources, Denver, Colorado, USA. Available: http://www.cnhp.colostate.edu/download/documents/2003/wetland_classification_final_report_2003.pdf.

Castelli, R. M., J. C. Chambers, and R. J. Tausch. 2000. Soil-plant relations along a soil-water gradient in great basin riparian meadows. Wetlands 20:251-266.

Caswell, H., editor. 2001. Matrix Population Models: Construction, Analysis, and Interpretation. Sinauer Associates, Inc., Sunderland, Massachusetts, USA.

Chadde, S. W., J. S. Shelly, R. J. Bursik, R. K. Moseley, A. G. Evenden, M. Mantas, F. Rabe, and B. Heidel. 1998. Peatlands on National Forests of the northern Rocky Mountains: ecology and conservation. USDA Forest Service Technical Report RMRS-GTR-11, Rocky Mountain Research Station, Fort Collins, Colorado, USA.

Chimner, R. A., and D. J. Cooper. 2003. Influence of water table levels on CO2 emissions in a Colorado subalpine fen: an in situ microcosm study. Soil Biology and Biochemistry 35:345-351.

Chimner, R. A., and D. J. Cooper. 2004. Using stable oxygen isotopes to quantify the water source used for transpiration by native shrubs in the San Luis Valley, Colorado USA. Plant and Soil 260:225-236.

Clearwater, M. J., F. C. Meinzer, J. L. Andrade, G. Goldstein and N. M. Holbrook. 1999. Potential errors in measurement of nonuniform sap flow using heat dissipation probes. Tree Physiology. 19:681-687.

Comer, P. S., S. Menard, M. Tuffly, K. Kindscher, R. Rondeau, G. Steinuaer, R. Schneider, and D. Ode. 2003. Upland and wetland ecological systems in Colorado, Wyoming, South Dakota, Nebraska, and Kansas. Report and map to the National Gap Analysis Program. NatureServe, Arlington, Virginia, USA.

Cooper, D. J., and R. E. Andrus. 1994. Patterns of vegetation and water chemistry in peatlands of the west-central Wind River Range, Wyoming, USA. Canadian Journal of Botany 72:1586-1597.

Cooper, D. J. 1996. Water and soil chemistry, floristics, and phytosociology of the extreme rich High Creek fen, in South Park, Colorado, U.S.A. Canadian Journal of Botany. 74: 1801-1811.

Cooper, D. J. and J. S. Sanderson. 1997. A montane *Kobresia myosuroides* fen community type in the Southern Rocky Mountains of Colorado, USA. Arctic and Alpine Research 29:300-303.

Cooper, D. J., L. H. MacDonald, S. K. Wenger, and S. W. Woods. 1998. Hydrologic restoration of a fen in Rocky Mountain National Park, Colorado, USA. Wetlands 18:335-345.

Cooper, D. J., D. M. Merritt, D. C. Andersen, and R. A. Chimner. 1999. Factors controlling the establishment of Fremont cottonwood seedlings on the upper Green River, U.S.A. Regulated Rivers: Research and Management 15:419-440.

Cooper, D.J., D. D'Amico, and M.L. Scott. 2003. Physiological and morphological response patterns of *Populus deltoides* to alluvial groundwater pumping. Environmental Management 31:215-226.

Cooper, D. J., R. Andrus, and C. D. Arp. 2002. *Sphagnum balticum* in a southern Rocky Mountain iron fen. Madrono 49:186-188.

Cooper, D. J., D. C. Andersen, and R. A. Chimner. 2003a. Multiple pathways for woody plant establishment on floodplains at local to regional scales. Journal of Ecology 91:182-196.

Cooper, D.J., D. D'Amico, and M.L. Scott. 2003b. Physiological and morphological response patterns of *Populus deltoides* to alluvial groundwater pumping. Environmental Management 31:215-226.

Cooper, D. J., J. Dickens, N. T. Hobbs, L. Christensen, and L. Landrum. 2006. Hydrologic, geomorphic and climatic processes controlling willow establishment in a montane ecosystem. Hydrological Processes 20:1845-1864.

Cooper, D. and E. Wolf. 2006. Yosemite Valley: Hydrologic Regime, Soils, Pre-Settlement vegetation, Disturbance and Concepts for Restoration. Unpublished Report to National Park Service, Yosemite National Park, CA.

Cooper, D. and L. Patterson. 2007. Restoration Plan for the Flagg Ranch, John D. Rockefeller Memorial Parkway, Wyoming. Unpublished Report to National Park Service, Grand Teton National Park, Moose, Wyoming.

Cooper, D. and E. Wolf. 2007. The influence of ground water pumping on wetlands in Crane Flat, Yosemite National Park, California. Unpublished Report to National Park Service, Yosemite National Park, CA.

Cooper, D.J., J.D. Lundquist, J. King, A. Flint, L. Flint, E. Wolf, and F.C. Lott. 2007. Effects of the Tioga Road on hydrologic processes and lodgepole invasion into Tuolumne Meadows, Yosemite National Park. Unpublished Report to National Park Service, Yosemite National Park, CA.

Cowardin, L. M., V. Carter, and E. T. La Roe. 1979. Classification of wetlands and deepwater habitats of the United States. FWS/OBS-79/31. USDI Fish and Wildlife Service, Washington, D.C., USA.

Crawley, M. J. 1986. Life history and environment. Pages 253-290 *in* M. J.Crawley, editor. Blackwell Scientific Publications, London and Boston.

Crum, H., and L. Anderson. 1981. Mosses of North America. Columbia University Press, New York, USA. Two volumes.

Davidson, A. S., and R. L. Knight. 2001. Avian nest success and community composition in a western riparian forest. Journal of Wildlife Management 65:334-344.

Disalvo, A. C., and S. C. Hart. 2002. Climatic and stream-flow controls on tree growth in a western montane riparian forest. Environmental Management 30:678-691.

Dixon, M. D. 2003. Effects of flow pattern on riparian seedling recruitment on sandbars in the Wisconsin River, Wisconsin, USA. Wetlands 23:125-139.

Dodd, J. D., and R. T. Coupland. 1966. Vegetation of saline areas in Saskatchewan. Ecology 47: 958-968.

Ehleringer, J. R., and C. B. Osmond. 1989. Stable isotopes. Pages 281-291 *in* R. W. Pearcy, J. Ehleringer, H. A. Mooney and P. W. Rundel, editors. 1989. Plant Physiological Ecology: field methods and instrumentation. Chapman and Hall, New York, USA.

Elzinga, C. L., D. W. Salzer, J. W. Willoughby, and J. P. Gibbs. 2001. Monitoring plant and animal populations. Blackwell Science, Inc., Malden, Massachusetts, USA.

FERC 1920. Federal Energy Regulatory Commission established under the Federal Power Act. Chapter 12 of Title 16 of the United States Code, entitled "Federal Regulation and Development of Power", Washington, D.C.

Friedman, J. M., and V. J. Lee. 2002. Extreme floods, channel change, and riparian forests along ephemeral streams. Ecological Monographs 72:409-425.

Fall, P. 1997. Fire history and composition of the subalpine forest of western Colorado during the Holocene. Journal of Biogeography 24:309-325.

Ferris, J. G. 1963. Cyclic water-level fluctuations as a basis for determining aquifer transmissibility. Pages 305-318 *in* R. Bentatl, R., editor. Methods of determining permeability, transmissibility and drawdown. USGS Water Supply Paper 1536-I, Washington, D.C., USA.

Francis, R. A., and A. M. Gurnell. 2006. Initial establishment of vegetative fragments within the active zone of a braided gravel-bed river (River Tagliamento, NE Italy). Wetlands 26:641-648.

Franz, E. H., and F. A. Bazzaz. 1977. Simulation of vegetation response to modified hydrologic regimes: a probablistic model based on niche differentiation in a floodplain forest. Ecology 58:176-183.

Friedman, J. M., and G. T. Auble. 1999. Mortality of riparian box elder from sediment mobilization and extended inundation. Regulated Rivers: Research and Management 15:463-476.

Friedman, J. M., G. T. Auble, E. D. Andrews, G. Kittel, R. F. Madole, E. R. Griffin, and T. M. Allred. 2006. Transverse and longitudinal variation in woody riparian vegetation along a montane river. Western North American Naturalist 66:78-91.

Friedman, J. M., and V. J. Lee. 2002. Extreme floods, channel change, and riparian forests along ephemeral streams. Ecological Monographs 72:409-425.

Graf, W. L. 1999. Dam nation: a geographic census of American dams and their large-scale hydrologic impacts. Water Resources Research 35:1305-1311

Gregory, S. V., F. J. Swanson, W. A. McKee, and K. W. Cummins. 1991. An ecosystem perspective of riparian zones. Bioscience 41:540-550.

Grime, J. P. 1977. Evidence for existence of 3 primary strategies in plants and its relevance to ecological and evolutionary theory. American Naturalist 111:1169-1194.

Grubb, P. J. 1977. The maintenance of species-richness in plant communities: the importance of the regeneration niche. Biological Reviews 52:107-145.

Harner, M. J., and J. A. Stanford. 2003. Differences in cottonwood growth between a losing and a gaining reach of an alluvial floodplain. Ecology 84:1453-1458.

Harrelson, C. C., C. L. Rawlins, and J. P. Potyondy. 1994. Stream channel reference sites: an illustrated guide to field technique. Gen.Tech. Rep. RM-245, Rocky Mountain Forest and Range Experiment Station, Fort Collins, Colorado, USA.

Haukos, D. A. 1992. Invertebrate herbivory in a Texas playa lake. Texas Journal of Science 44: 254-257.

Heidel, B., and S. Laursen. 2003. Botanical and ecological inventory of peatland sites on the Medicine Bow National Forest. Wyoming Natural Diversity Database, Laramie, Wyoming, USA.

Horton, J. L., and J. L. Clark. 2001. Water table decline alters growth and survival of *Salix gooddingii* and *Tamarix chinensis* seedlings. Forest Ecology and Management 140:239-247.

Horton, J. S., F. C. Mounts, and J. M. Kraft. 1960. Seed germination and seedling establishment of phreatophyte species. USDA Forest Service Station Paper 48, Rocky Mountain Forest and Range Experiment Station, Fort Collins, Colorado, USA.

Houle, G. 1994. Spatiotemporal patterns in the components of regeneration of four sympatric tree species--*Acer rubrum, A. saccharum, Betula alleghaniensis and Fagus grandifolia.* Journal of Ecology 82:39-53.

Hubbard, R. M., J. Stape, M.G. Ryan, A. Almeida, and J. Rojas. 2010. Effects of irrigation on water use and water use efficiency in two fast growing eucalyptus plantations. Forest Ecology and Management 259:1714-1721.

Hughes, F. M. R., T. Harris, K. Richards, G. Pautou, A. F. L. Hames, N. Barsoum, J. Girel, J. L. Peiry, and R. Foussadier. 1997. Woody riparian species response to different soil moisture conditions: laboratory experiments on *Alnus incana* (L.) *Moench*. Global Ecology and Biography Letters 6:247-256.

Hutchinson, G. E. 1957. Concluding remarks. Cold Spring Harbor Symposia on Quantitative Biology 22:415–427.

Jackson, P., F. Meinzer, G. Goldstein, N. Holbrook, J. Colvelier, and F. Rada. 1993. Environmental and physiological influences on carbon isotope composition of gap and understory plants in a lowland tropical forest. Chapter 9 *in* J. Ehleringer, A. Hall, and G. Farquhar, editors. Stable isotopes and plant carbon-water relations. Academic Press, New York, USA.

Jansson, R., C. Nilsson, M. Dynesius, and E. Andersson. 2000. Effects of river regulation on river-margin vegetation: a comparison of eight boreal rivers. Ecological Applications 10:203-224.

Johnson, W. C. 2000. Tree recruitment and survival in rivers: influence of hydrological processes. Hydrological Processes 14:3051-3074.

Johnson, W. C., T. L. Sharik, R. A. Mayes, and E. P. Smith. 1987. Nature and cause of zonation discreteness around glacial prairie marshes. Canadian Journal of Botany 65:1622-1632.

Jolly, I. D., G. R. Walker, and P. J. Thorburn. 1993. Salt accumulation in semiarid floodplain soils with implications for forest health. Journal of Hydrology 150:589-614.

Jongman, R., C. ter Braak, and O. Van Tongeren, editors. 1995. Data Analysis in Community and Landscape Ecology. Second edition. Cambridge University Press, Cambridge, United Kingdom.

Karrenberg, S., P. J. Edwards, and J. Kollmann. 2002. The life history of *Salicaceae* living in the active zone of floodplains. Freshwater Biology 47:733-748.

Kalra, A., T. C. Piechota, R. Davies, and G. A. Tootle. 2008. Changes in US streamflow and western US snowpack. Journal of Hydrologic Engineering 13:156-163.

Knighton, D. 1998. Fluvial Forms and Processes--A New Perspective. Arnold Publishing Co., London, United Kingdom.

Lambeck, R. J. 1997. Focal species: a multi-species umbrella for nature conservation. Conservation Biology 11:849-856.

Legendre, P., and L. Legendre, editors. 1998. Numerical Ecology. Second edition. Elsevier, Amsterdam, Netherlands.

Lenssen, J., F. Menting, W. Van Der Putten, and K. Blom. 1999. Control of plant species richness and zonation of functional groups along a freshwater flooding gradient. Oikos 86:523-534.

Lichvar, R. W. and J. T. Kartesz. 2012. North American Digital Flora: National Wetland Plant List, version 3.0 (http://wetland_plants.usace.army.mil).

Lins, H. F. 1997. Regional streamflow regimes and hydroclimatology. Water Resources Research 33:1655-1667.

Lite, S. J., K. J. Bagstad, and J. C. Stromberg. 2005. Riparian plant species richness along lateral and longitudinal gradients of water stress and flood disturbance, San Pedro River, Arizona, USA. Journal of Arid Environments 63:785-813.

Lite, S. J. and J. C. Stromberg. 2005. Surface water and ground-water thresholds for maintaining Populus-Salix forests, San Pedro river, Arizona. Biological Conservation 125: 153-167.

Lytle, D. A., and D. M. Merritt. 2004. Hydrologic regimes and riparian forests: a structured population model for cottonwood. Ecology 85:2493-2503.

Mahoney, J. M., and S. B. Rood. 1998. Streamflow requirements for cottonwood seedling recruitment--an integrative model. Wetlands 18:634-645.

Manly, B. F. J., editor. 1994. Multivariate Statistical Methods: A Primer. Second edition. Chapman and Hall, London.

McClymont, A. F., J. W. Roy, M. Hayashi, L. R. Bentley, H. Maurer, and G. Langston. 2011. Investigating groundwater flow paths within proglacial moraine using multiple geophysical methods. Journal of Hydrology 399:57-69.

McCune, B., and J. B. Grace. 2002. Analysis of Ecological Communities. MJM Press, Gleneden Beach, Oregon, USA. 304 pages.

McCune, B, and M. J. Mefford. 1999. PC-ORD, Multivariate analysis of ecological data. Fourth edition. MjM Software Design, Gleneden Beach, Oregon, USA.

Merigliano, M. F. 2005. Cottonwood understory zonation and its relation to floodplain stratigraphy. Wetlands 25:356-374.

Merritt, D. M., and D. J. Cooper. 2000. Riparian vegetation and channel change in response to river regulation: a comparative study of regulated and unregulated streams in the Green River basin, USA. Regulated Rivers: Research and Management 16:543-564.

USDA Forest Service Gen. Tech. Rep. RMRS-GTR-282. 2012

121

Merritt, D. M., M. L. Scott, N. L. Poff, G. T. Auble, and D. A. Lytle. 2010a. Theory, methods and tools for determining environmental flows for riparian vegetation: riparian vegetation-flow response guilds. Freshwater Biology 55:206-225.

Merritt, D. M., H. L. Bateman, and C. D. Peltz. 2010b. Instream flow requirements for maintenance of wildlife habitat and riparian vegetation: Cherry Creek, Tonto National Forest, Arizona. A report submitted to the Tonto National Forest by USDA Forest Service Stream Systems Technology Center, Fort Collins, Colorado, USA.

Merritt, D. M., and N. L. Poff. 2010. Shifting dominance of riparian *Populus* and *Tamarix* along gradients of flow alteration in western North American rivers. Ecological Applications 20:135-152.

Mills, G. C., J. B. Dunning, Jr., and J. M. Bates. 1991. The relationship between breeding bird density and foliage volume. Wilson Bulletin 103:468-479.

Mitsch, W. J., and J. G. Gosselink. 2000. Wetlands. Third edition. John Wiley, New York, USA.

Montgomery, D. M. 1999. Process domains and the river continuum. Journal of the American Water Resources Association 35:397-410.

Mueller-Dombois, D., and H. Ellenberg. 1974. Aims and Methods of Vegetation Ecology. The Blackburn Press, Caldwell, New Jersey, USA.

Naiman, R. J., and H. Decamps. 1997. The ecology of interfaces: riparian zones. Annual Review of Ecology and Systematics 28:621-658.

Naiman, R. J., H. DeCamps, and M. Pollack. 1993. The role of riparian corridors in maintaining regional biodiversity. Ecological Applications 3:209-212.

National Research Council. 1995. Wetlands: Characteristics and Boundaries. National Academy Press, Washington, DC, USA.

National Research Council. 2002. Riparian Areas: Functions and Strategies for Management. National Academy Press, Washington DC, USA

Naumburg, E., R. Mata-Gonzalez, R.G. Hunter, T. Mclendon, and D.W. Martin. 2005. Phreatophytic vegetation and groundwater fluctuations: a review of current research and application of ecosystem response modeling with an emphasis on Great Basin vegetation. Environmental Management 35:726-740.

Nelson, R. W., W. J. Logan, and E. C. Weller. 1984. Playa wetlands and wildlife on the Southern Great Plains: A characterization of habitat. U.S. Environmental Protection Agency, Washington, DC, USA.

Nilsson, C., A. Ekblad, M. Gardfjell, and B. Carlberg. 1991. Long-term effects of river regulation on river margin vegetation. Journal of Applied Ecology 28:963-987.

Nilsson, C., A. Ekblad, M. Dynesius, S. Backe, M. Gardfjell, B. Carlberg, S. Hellqvist, and R. Jansson. 1994. A comparison of species richness and traits of riparian plants between a main channel and its tributaries. Journal of Ecology 82:281-295.

Nilsson, C., and M. Svedmark. 2002. Basic principles and ecological consequences of changing water regimes: riparian plant communities. Environmental Management 30:468-480.

Novitzki, R. 1978. Hydrologic characteristics of Wisconsin wetlands, and their influence on floods, stream flow and sediment. Pages 377-388, in Wetland functions and values: the state of our understanding. American Water Resources Association.

Novitski, R.P 1982. Hydrology of Wisconsin Wetlands. University of Wisconsin Extension Geological Natural History Survey Circular 40, University of Wisconsin, Madison, 22 pp.

Olden, J. D., and N. Poff, L. 2003. Redundancy and the choice of hydrologic indices for characterizing streamflow regimes. River Research and Applications 19:101-121.

Oren, R., J. S. Sperry, et al. 1999. Survey and synthesis of intra- and interspecific variation in stomatal sensitivity to vapour pressure deficit. Plant Cell and Environment 22(12):1515-1526.

Palmer, W. C. 1965. Meteorological drought. U.S. Weather Bureau Research Paper No. 45, Washington, DC, USA.

Patten, D. T. 1998. Riparian ecosystems of semi-arid North America: diversity and human impacts. Wetlands 18:498-512.

Pearcy, R. W., J. Ehleringer, H. A. Mooney and P. W. Rundel, editors. 1989. Plant Physiological Ecology: field methods and instrumentation. Chapman and Hall, New York, USA.

Pinay, G., J. C. Clement, R.J. Naiman. 2002. Basic principles and ecological consequences of changing water regimes on nitrogen cycling in fluvial systems. Environmental Management 30:481-491.

Pinay, G., B. Gumiero, E. Tabacchi, O. Gimenez, A. M. Tabacchi-Planty, M. M. Hefting, T. P. Burt, V. A. Black, C. Nilsson, V. Iordache, F. Bureau, L. Vought, G. E. Petts, and H. Décamps. 2007. Patterns of denitrification rates in European alluvial soils under various hydrological regimes. Freshwater Biology 52:252-266.

Pinay, G., and R. J. Naiman. 1991. Short-term hydrologic variations and nitrogen dynamics in beaver created meadows. Archiv Fur Hydrobiologie 123:187-205.

Pockman, W. T. and J. S. Sperry. 2000. Vulnerability to xylem cavitation and the distribution of Sonoran desert vegetation. American Journal of Botany 87:1287-1299.

Pollock, M. M. 1998. Biodiversity. Page 705 *in* R. J. Naiman and R. E. Bilby, editors. River Ecology and Management Lessons from the Pacific Coast Eco-region. Springer-Verlag, New York, USA.

Poff, N. L. 1996. A hydrogeography of unregulated streams in the United States and an examination of scale-dependence in some hydrological descriptors. Freshwater biology 36:71-91.

Pringle, C. M. 2000. Threats to US public lands from cumulative hydrologic alterations outside of their boundaries. Ecological Applications 10:971-989.

Rains, M. C., J. E. Mount, and E. W. Larsen. 2004. Simulated changes in shallow groundwater and vegetation distributions under different reservoir operations scenarios. Ecological Applications 14:192-207.

Rathburn, S. L., D. M. Merritt, E. E. Wohl, J. A. Sanderson, and H. A. L. Knight. 2009. Characterizing environmental flows for maintenance of river ecosystems: North Fork Cache La Poudre River, Colorado. Pages 143-157 *in* L. A. James, S. L. Rathburn, and G. R. Whittecar, editors. Management and restoration of fluvial systems with broad historical changes and human impacts. Geological Society of America Special Paper, 451.

Raunkiær, C. 1934. The life forms of plants and statistical plant geography. Oxford University Press, Oxford, United Kingdom.

Reed, P. B. 1988. National list of plant species that occur in wetlands: national summary. U.S. Fish and Wildlife Service Biological Report 88. U.S. Fish and Wildlife Service, Washington, DC, USA.

Renofalt, B. M., D. M. Merritt, and C. Nilsson. 2007. Connecting variation in vegetation and stream flow: the role of geomorphic context in vegetation response to large floods along boreal rivers. Journal of Applied Ecology 44:147-157.

Richter, B. D., J. V. Baumgartner, R. Wigington, and D. P. Braun. 1997. How much water does a river need? Freshwater Biology 37:231-249.

Richter, B. D., R. Mathews, and R. Wigington. 2003. Ecologically sustainable water management: managing river flows for ecological integrity. Ecological Applications 13:206-224.

Rood, S. B., L. A. Goater, J. M. Mahoney, C. M. Pearce, and D. G. Smith. 2007. Floods, fire, and ice: disturbance ecology of riparian cottonwoods. Canadian Journal of Botany--Revue Canadienne De Botanique 85:1019-1032.

Rood, S. B., C. R. Gourley, E. M. Ammon, L. G. Heki, J. R. Klotz, M. L. Morrison, D. Mosley, G. G. Scoppettone, S. Swanson, and P. L. Wagner. 2003. Flows for floodplain forests: a successful riparian restoration. Bioscience 53:647-656.

Rood, S., S. Patino, K. Coombs, and M. Tyree. 2000. Branch sacrifice: cavitation-associated drought adaptation of riparian cottonwoods. Trees 14:248-257.

Rood, S. B., G. M. Samuelson, J. H. Braatne, C. R. Gourley, F. M. R. Hughes, and J. M. Mahoney. 2005. Managing river flows to restore floodplain forests. Frontiers in Ecology and the Environment 3:193-201.

Scott, M. L., G. T. Auble, and J. M. Friedman. 1997. Flood dependency of cottonwood establishment along the Missouri River, Montana, USA. Ecological Applications 7:677-690.

Scott, M. L., J. M. Friedman, and G. T. Auble. 1996. Fluvial process and the establishment of bottomland trees. Geomorphology 14:327-339.

Scott, M. L., P. B. Shafroth, and G. T. Auble. 1999. Responses of riparian cottonwoods to alluvial water table declines. Environmental Management 23:347-358.

Scott, M. L., M. A. Wondzell, and G. T. Auble. 1993. Hydrograph characteristics relevant to the establishment and growth of western riparian vegetation: 237-246. Hydrology Days Publications, Atherton, California, USA.

Seabloom, E. W., A. G. Van Der Valk, and K. A. Moloney. 1998. The role of water depth and soil temperature in determining initial composition of prairie wetland coenoclines. Plant Ecology 138:203-216.

Segelquist, C.A., M. L. Scott, and G. T. Auble. 1993. Establishment of *Populus deltoides* under simulated alluvial groundwater declines. American Midland Naturalist 130:274-285.

Shafroth, P. B., G. T. Auble, J. C. Stromberg, and D. T. Pattern. 1998. Establishment of woody riparian vegetaion in relation to annual patterns of streamflow, Bill Williams River, Arizona. Wetlands 18:577-590.

Shaw, J., and D. J. Cooper. 2008. Watershed and stream reach characteristics controlling riparian vegetation in semiarid ephemeral stream networks. Journal of Hydrology 350:68-82.

Shipley, B., P. A. Keddy, and L. P. Lefkovitch. 1991. Mechanisms producing plant zonation along a water depth gradient: a comparison with the exposure gradient. Canadian Journal of Botany 69:1420-1424.

Smith, L. M., and D. A. Haukos. 2002. Floral diversity in relation to playa wetland area and watershed disturbance. Conservation Biology 16:964-974.

Smith, L. M., and J. A. Kadlec. 1983. Seed banks and their role during drawdown of a North American marsh. Journal of Applied Ecology 20:673-684

Smith, R. D., A. Ammann, C. Bartoldus, and M. M. Brinson. 1995. An approach for assessing wetland functions using hydrogeomorphic classification, reference wetlands, and functional indices. United States Army Corps of Engineers Technical Report WRP-DE-9. U.S. Army Engineer Waterways Experiment Station, Vicksburg, Mississippi, USA.

Smith, S. D., D. A. Devitt, A. Sala, J. R. Cleverly, and D. E. Busch. 1998. Water relations of riparian plants from warm desert regions. Wetlands 18:687-696.

Sperry, J. S. 1995. Limitations on stem water transport and their consequences. Pages ??? *in* B. L. Gartner editor. Plant stems:physiology and functional morphology. Academic Press, San Diego, USA.

Sperry, J. S. 2000. Hydraulic constraints on plant gas exchange. Agricultural and Forest Meteorology 104:13-23.

Steppe, K., D. DePauw, T. Doody, and R. O. Teskey. 2010. A comparison of sap flux density using thermal dissipation, heat pulse velocity and heat field deformation methods. Agricultural and Forest Meteorology 150:1046-1056.

Stromberg, J. C., V. B. Beauchamp, M. D. Dixon, S. J. Lite and C. Paradzick. 2007a. Importance of low-flow and high-flow characteristics to restoration of riparian vegetation along rivers in and south-western United States. Freshwater Biology 52:651-679.

Stromberg, J. C., S. J. Lite, R. Marler, C. Paradzick, P. B. Shafroth, D. Shorrock, J. M. White, and M. S. White. 2007b. Altered stream-flow regimes and invasive plant species: the *Tamarix* case. Global Ecology and Biogeography 16:381-393.

Stromberg, J. C., and D. T. Patten. 1990. Riparian vegetation instream flow requirements: a case study from a diverted stream in the eastern Sierra Nevada, California, USA. Environmental Management 14:185-194.

Stromberg, J. C., and D. T. Patten. 1991. Instream flow requirements for cottonwoods at Bishop Creek, Inyo County, California, USA. Rivers 2:1-11.

Stromberg, J. C., and D. T. Patten. 1996. Instream flow and cottonwood growth in the eastern Sierra Nevada of California, USA. Regulated Rivers: Research and Management 12:1-12.

Stromberg, J.C., B. D. Richter, D. T. Patten, and L.G. Wolden. 1993. Response of a Sonoran riparian forest to a 10-year return flood. Great Basin Naturalist 53:118-130.

Szaro, R. C. 1990. Southwestern riparian plant communities: site characteristics, tree species distributions, and size-class structures. Forest Ecology and Management 33/34:315-334.

Taneda, H., and J. S. Sperry. 2008. A case-study of water transport in co-occurring ring- versus diffuse-porous trees: contrasts in water-status, conducting capacity, cavitation and vessel refilling. Tree Physiology 28:1641-1651.

Thormann, M. N., A. R. Szumigalski, and S. E. Bayley. 1999. Aboveground peat and carbon accumulation potentials along a bog-fen-marsh wetland gradient in southern boreal Alberta, Canada. Wetlands 19:305-317.

Tiner, R. W. 1984. Wetlands of the United States: current status and recent trends. USDI Fish and Wildlife Service, National Wetlands Inventory, Washington, DC, USA.

Tyree, M., and J. Sperry. 1989. Vulnerability of xylem to cavitation and embolism. Annual Review of Plant Physiology and Plant Molecular Biology 40:19-36.

Tyree, M. T., K. J. Kolb, S. J. Rood, and S. Patino. 1994. Vulnerability to drought induced cavitation of riparian cottonwoods in Alberta: a possible factor in the decline of the ecosystem? Tree Physiology 14:455-466.

Ungar, I. A. 1966. Salt tolerance of plants growing in saline areas of Kansas and Oklahoma. Ecology 47:154-162.

Ungar, I. A. 1974. Halophyte communities of Park County, Colorado. Bulletin of the Torrey Botanical Club 101:145-152.

U.S. Army Corps of Engineers. 2009. Mountains and plains regional supplement to the Corps of Engineers wetland delineation manual: western mountains, valleys, and coast region (Version 2.0). U.S. Army Engineer Research and Development Center Environmental Laboratory, Vicksburg, Mississippi, USA.

Van der Valk, A. G., and C. B. Davis. 1976. Seed banks of prairie glacial marshes. Canadian Journal of Botany 54:1832-1838.

Van der Valk, A.G., L. Squires, C.H. Welling. 1994. Assessing the impacts of an increase in water-level on wetland vegetation. Ecological Applications 4:525-534.

Vanderijt, C., L. Hazelhoff, and C. Blom. 1996. Vegetation zonation in a former tidal area: a vegetation-type response model based on DCA and logistic regression using GIS. Journal of Vegetation Science 7:505-518.

Waddle, T. J., and K. D. Bovee. 2009. Environmental flows studies of the Fort Collins Science Center, U.S. Geological Survey: Cherry Creek, Arizona. U.S. Geological Survey Reston, Virginia, USA.

Wafer, C. C., J. B. Richards, and D. L. Osmond. 2004. Construction of platinum-tipped redox probes for determining soil redox potential. Journal of Environmental Quality 33:2375-2379.

Weber, W. A., and R. Wittmann. 2001. Colorado flora, western slope. Colorado Associated University Press, Niwot, Colorado, USA.

Weltzin, J. F., J. Pastor, C. Harth, S. D. Bridgham, K. Updegraff, and C. T. Chapin. 2000. Response of bog and fen plant communities to warming and water-table manipulations. Ecology 81:3464-3478.

Westbrook, C., D. J. Cooper, B. Baker. 2006. Beaver dams and overbank floods influence groundwater-surface water interactions of a Rocky Mountain riparian area. Water Resources Research 42:W06404. doi:10.1029/2005WR004560.

Williams, C. A. and D. J. Cooper. 2005. Mechanisms of riparian cottonwood decline along regulated rivers. Ecosystems 8:382-395.

Willms, J., S. B. Rood, W. Willms, and M. Tyree. 1998. Branch growth of riparian cottonwoods: a hydrologically sensitive dendrochronological tool. Trees--Structure and Function 12:215-223.

Wilson, S. D., D. R. J. Moore, and P. A. Keddy. 1993. Relationships of marsh seed banks to vegetation patterns along environmental gradients. Freshwater Biology 29:361-370.

Winter, T. C., J. W. Harvery, O. L. Franke, and W. M. Alley. 1996. Ground water and surface water, a simple resource. USGS Circular 1139.

Winter, T. C., 1989. Hydrologic studies of wetlands in the northern prairie. Pages 16-54 *in* A. Van Der Valk, editor. Northern Prairie Wetlands. Iowa State University Press, Ames, Iowa, USA

Winter, T. C. 1999. Relation of Streams, Lakes, and Wetlands to Groundwater Flow Systems. Hydrogeology Journal 7:28-45.

Winter, T. C. 2001. The concept of hydrologic landscapes. Journal of the American Water Resources Association 37:335-349.

Winter, T. C., and D. O. Rosenberry. 1998. Hydrology of prairie pothole wetlands during drought and deluge: a 17-year study of the Cottonwood Lake wetland complex in North Dakota in the perspective of longer term measured and proxy hydrological records. Climatic Change 40:189-209.

Winter, T. C., D. O. Rosenberry, D. C. Buso, and D. A. Merk. 2001. Water source to four U.S. wetlands: implications for wetland management. Wetlands 21: 462-473.

Wohl, E. E. 2000. Mountain Rivers. American Geophysical Union, Washington, DC, USA.

Wurster, F. C., D. J. Cooper, and W. E. Sanford. 2003. Stream/aquifer interactions at Great Sand Dunes National Monument, Colorado: influences on interdunal wetland disappearance. Journal of Hydrology 271:77-100.

The Rocky Mountain Research Station develops scientific information and technology to improve management, protection, and use of the forests and rangelands. Research is designed to meet the needs of the National Forest managers, Federal and State agencies, public and private organizations, academic institutions, industry, and individuals. Studies accelerate solutions to problems involving ecosystems, range, forests, water, recreation, fire, resource inventory, land reclamation, community sustainability, forest engineering technology, multiple use economics, wildlife and fish habitat, and forest insects and diseases. Studies are conducted cooperatively, and applications may be found worldwide.

Station Headquarters
Rocky Mountain Research Station
240 W Prospect Road
Fort Collins, CO 80526
(970) 498-1100

Research Locations

Flagstaff, Arizona	Reno, Nevada
Fort Collins, Colorado	Albuquerque, New Mexico
Boise, Idaho	Rapid City, South Dakota
Moscow, Idaho	Logan, Utah
Bozeman, Montana	Ogden, Utah
Missoula, Montana	Provo, Utah

www.fs.fed.us/rmrs